★

Deep Time
and the Texas High Plains

Grover E. Murray Studies
in the American Southwest

ALSO IN THE GROVER E. MURRAY STUDIES IN THE AMERICAN SOUTHWEST

Cacti of the Trans-Pecos and Adjacent Areas
A. Michael Powell and James F. Weedin

DEEP TIME

and the

Texas High Plains

History and Geology

Paul H. Carlson

Texas Tech University Press

This book is typeset in Adobe Garamond. The paper used in this book meets the minimum requirements of ANSI/NISO Z39.48-1992 (R1997).

LIBRARY OF CONGRESS CATALOGING-IN-PUBLICATION DATA

Carlson, Paul H.

Deep time and the Lubbock Lake landmark : history and geology in the upper Brazos River region / Paul H. Carlson.

p. cm. — (Grover E. Murray studies in the American Southwest)

Summary: "Surveys the history and geologic past of the Texas High Plains and upper Brazos River region by focusing on human activity and adaptation and on shifting environmental conditions and animal resources on the Llano Estacado and in Yellow House Draw, the site of the current Lubbock Lake Landmark"—Provided by publisher.

Includes bibliographical references and index.

ISBN-13: 978-0-89672-552-2 (cloth : alk. paper)

ISBN-10: 0-89672-552-9 (cloth : alk. paper)

ISBN-13: 978-0-89672-553-9 (pbk. : alk. paper)

ISBN-10: 0-89672-553-7 (pbk. : alk. paper) 1. Lubbock Lake Site (Tex.) 2. Excavations (Archaeology)—Texas—Brazos River Region. 3. Excavations (Archaeology)—High Plains (U.S.) 4. Paleo-Indians—Texas—Brazos River Region. 5. Paleo-Indians—Texas—High Plains (U.S.) 6. Brazos River Region (Tex.)—Antiquities. 7. High Plains (U.S.)—Antiquities. 8. Natural history—Texas—Brazos River Region. 9. Natural history—High Plains (U.S.) I. Title. II. Series.

E78.T4C368 2005

917.64'01—dc22

Printed in the United States of America

05 06 07 08 09 10 11 12 13 / 9 8 7 6 5 4 3 2 1

SB

Texas Tech University Press

Box 41037

Lubbock, Texas 79409-1037 USA

800.832.4042

ttup@ttu.edu

www.ttup.ttu.edu

This book is dedicated to the memory of
William Curry Holden
and to the memory of four influential college professors
J. Leonard Jennewein
William J. Hughes
Cyril Allen
Ernest Wallace

★

Contents

★

Illustrations

Figures and Tables

Photographs

★

Preface

THIS BOOK grew from a series of visits to the Lubbock Lake Landmark, the natural history and archaeological site that Texas Tech University controls. With each visit my interest in the site's long past grew. I wanted to know more about geological forces that created the Llano Estacado, about human groups who had occupied and used the Lubbock Lake area and related archaeological sites, and about the animals that humans had pursued in the area. I wanted a history of the place.

In the visitor's center gift shop at the Landmark I found only a few books that might help. In an indirect way, several volumes related to the site, to the Llano Estacado, or to the upper Brazos River region. In a more direct way, the gift shop contained some highly specialized studies, a technical book on late Quaternary studies, and some children's works, including a coloring book. Because no broad, general account existed, not even a brief one, I determined, as an effort to learn more about the Lubbock Lake site, to try my hand at a short history, one designed for an educated reading public.

Accordingly, this book is an effort to review clearly and succinctly the fascinating record of intermittent human activity at the Lubbock Lake site and by extension a small portion of the larger Llano Estacado, or Texas High Plains. It is also an effort to examine briefly the area's long geological past.

My purpose, then, was to go back through what John McPhee has called "deep time." By doing so, I hoped to show in a concise fashion how the North American continent reached its present location, to survey the development of life on Earth, and to summarize the forces that created the

Texas High Plains and the Lubbock Lake site. Most of all, however, I wanted to provide a history of human activity in the Lubbock Lake area from the Paleoindian and Archaic periods through the Modern era.

The work is history. Although the reports of geologists, archaeologists, and other scientists figure prominently in it, the book in its methodology, style, and form is history.

As the bibliography makes clear, I am indebted to many people from many scholarly disciplines who have written about the topics covered in this study. Especially important for the Lubbock Lake site and the Texas High Plains are the books and reports of Eileen Johnson of Texas Tech University and Vance T. Holliday of the University of Arizona.

I am also indebted to the College of Arts and Sciences at Texas Tech University. In the fall of 2003 it granted me a semester-long research and writing leave, which I used to finish the manuscript. Relatedly, I owe great thanks to Bruce C. Daniels, former chair of the History Department at Texas Tech. Without his support the project could not have been completed in a timely fashion; indeed, it might not have been finished at all.

I owe thanks to others. At the Texas Tech University Library Jack Becker, John Hufford, and Eric Law especially were helpful. At the Lubbock Lake Landmark, Sue Shores and Terri Carnes offered information and suggestions. In their always efficient and courteous ways Debbie Shelfer and Peggy Ariaz in Tech's History Department provided assistance. Judith Keeling and Noel Parsons at Texas Tech University Press gave encouragement and guidance. Bryan Edwards, Joanne Dickenson, David J. Murrah, Stanley Marsh 3, Robert Hall, Leland Turner, M. Todd Houck, Damon Kennedy, Jeff Lee, Bo Brown, Sally Abbe, and Pam Wille provided assistance, information, or encouragement.

Personnel at Texas Tech University's fine Southwest Collection as always were competent, courteous, and full of good advice. William Tydeman and Tai Kreidler offered help. Jennifer Spurrier provided a study carrel so that I might leave many of the technical and scientific reports out and open while I was absent from the Collection's reading room. Patt Perry, with her usual good cheer, hard work, and pleasant conversation, made researching easier than it might have been. Janet Neugebauer and Fredonia Paschall helped with photos and rare documents. Their student assistants, especially Randy Vance, Christopher Taylor, M'Lyne Niehues, and Ben Vasquez, aided computer searches, retrieved books and scientific reports, made document copies, and otherwise provided thoughtful assistance.

Monte L. Monroe, the Southwest Collection's archivist, helped in many ways. He located obscure materials, suggested ideas, searched through uncataloged collections, and asked endless questions, always pushing me to complete the final document. Matt Crawford, Nora Mata, and Lindsay Starr prepared the maps, charts, and other illustrations.

Several people read early versions of the study. Bobbie Speer at West Texas A&M University and Frederick W. Rathjen, one of the Texas Panhandle's most accomplished historians, read an early version of some chapters, suggesting changes that eliminated pitfalls. David La Vere at the University of North Carolina at Wilmington read a draft of the manuscript and offered much good advice. Helen Tanner offered much good advice and was especially helpful with ideas about the maps. Eddie Guffee of the Wayland Baptist University Archaeological Laboratory read a later version of the manuscript. I am deeply grateful to everyone who helped, but, of course, I alone accept responsibility for any errors or misinterpretations that might exist.

Once again my wife, Ellen, throughout the long research, writing, and editing process provided support of the highest order. It was in fact after several months of research and during a long automobile trip that our conversation about the project led to the manuscript's basic form and final outline.

★

Deep Time
and the Texas High Plains

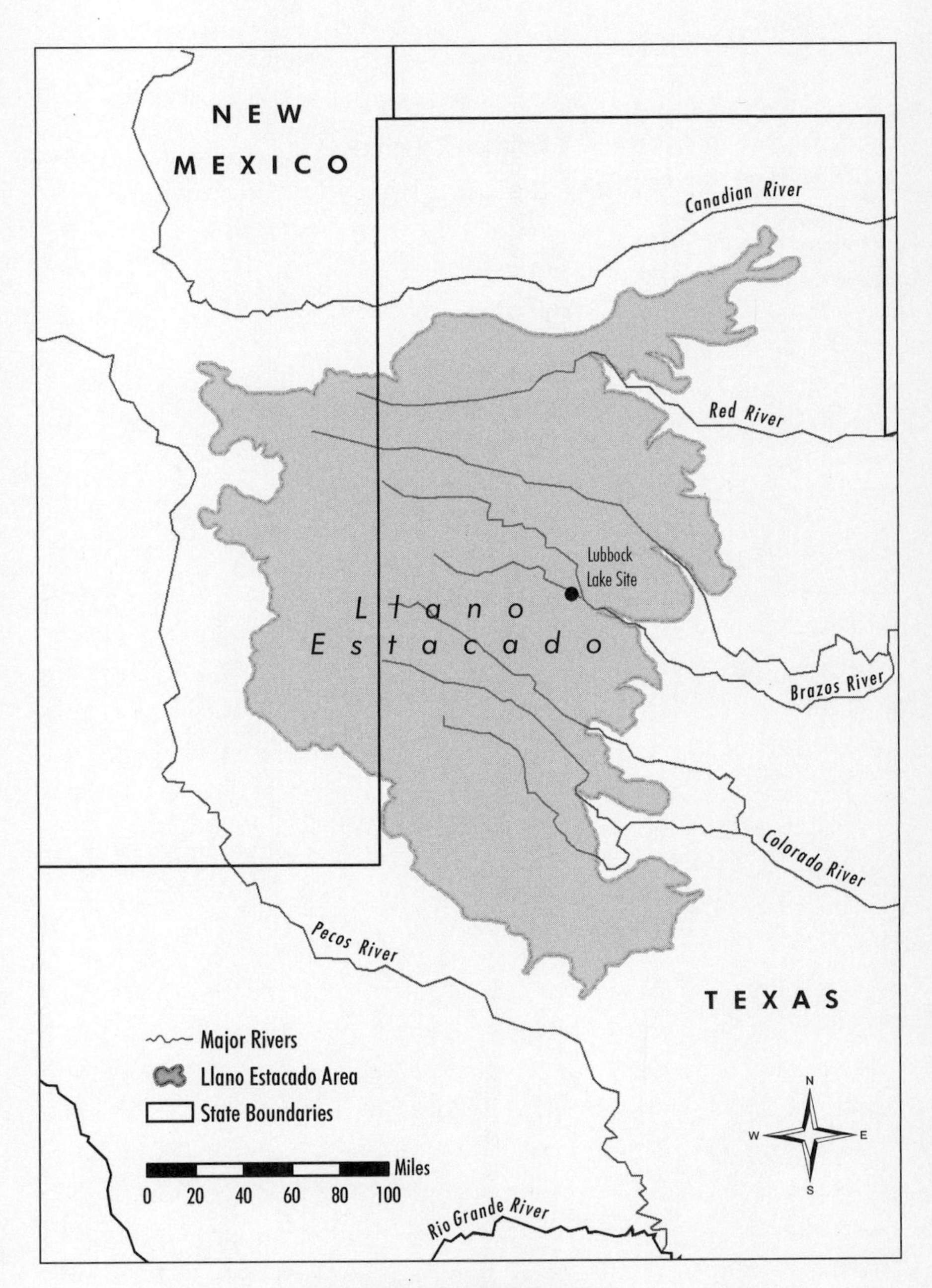

The Llano Estacado

★

Chapter 1

Constructing Deep Time

Putting Down Dirt

ARCHAEOLOGICAL investigations at the Lubbock Lake site, the Blackwater Draw sites, and other places on the Texas High Plains, including Running Water Draw, suggest that humans have occupied the Llano Estacado of western Texas and eastern New Mexico for nearly twelve thousand years. Such an extended, but not continuous, habitation means that people have occupied the Llano Estacado country for almost as long as humans have lived in the western hemisphere. In addition, fossil records suggest that dinosaurs and other prehistoric creatures once lived on land or in waters buried deep beneath the Llano Estacado's modern surface. Indeed, the ancient animals represent some of the oldest dinosaurs in Texas. The story reaches back through deep time to the moment of creation.[1]

It started with the big bang. Between 10 billion and 15 billion years ago, tiny, subatomic particles such as neutrinos, neutrons, and quarks, in a massive explosion, burst with enormous force and heat from a dense spot and in less than an instant raced across a dark, cold, endless space. It was chaos, but quickly gravitational, electromagnetic, and other natural forces brought order to the expanding cosmos. The universe had begun. And over time changes came. Particles produced atoms and the atoms formed elements. Gases also formed. Eventually, the gases and particle dust swirled into billions of huge galaxies. Stars evolved. The stars, some 10 bil-

Era/Subera	Period/Subperiod	Epoch	Millions of Years Ago
Cenozoic	Quarternary	Holocene	.01
		Pleistocene	1.6
Tertiary	Neogene	Pliocene	5.1
		Miocene	24
	Paleogene	Oligocene	38
		Eocene	55
		Paleocene	65
Mesozoic	Cretaceous		
	Jurassic		
	Triassic		
Paleozoic	Permian		286
	Carboniferous	Pennsylvanian	320
		Mississippian	360
	Devonian		408
	Silurian		438
	Ordovician		505
	Cambrian		570
	Precambrian		

Geologic Time Scale

Adapted from several sources; see particularly Martin Lockley and Adrian P. Hunt, *Dinosaur Tracks and Other Footprints of the Western United States* (New York: Columbia University Press, 1998), 8; or James Lawrence Powell, *Mysteries of Terra Firma* (New York: The Free Press, 2001), 11.

lion trillion of them, brought light and heat and attracted planets. In turn, the planets attracted moons and other galactic phenomena, such as asteroids, comets, and meteorites.

Time passed. Billions of years passed. Then, approximately 4.6 billion years ago on the distant edge of a spiral arm of the Milky Way, our own massive galaxy, interstellar gas and dust condensed to form Earth. The laws of physics—which are the same throughout the universe—hold Earth in orbit around the sun, one of 400 billion stars that make up the

Milky Way. Earth spins, of course, as it circles the sun, and as part of the Milky Way, it travels through space.

This large space traveler is more than 7,900 miles in diameter at the equator and some 24,900 miles in circumference. Its outer crust varies from five to twenty miles deep, with its thinner parts located on the ocean floors. Below the crust, where temperatures are hot enough to melt rock, is a deep mantle of material formed from silicon, aluminum, iron, magnesium, and other elements from which rocks form. Below the mantle, which is about 1,800 miles thick, is the Earth's core. The outer core and the inter mantle represent a hot, molten, sluggish mass that churns and twists in great loops and sends molten rock, or magma, toward the crust. If the magma breaks through the Earth's crust, volcanoes occur. More often, however, as it nears the crust, the hot mass cools to form hard rocks or sinks back into the mantle.

The rising and sinking of the magma has altered the Earth's surface. It

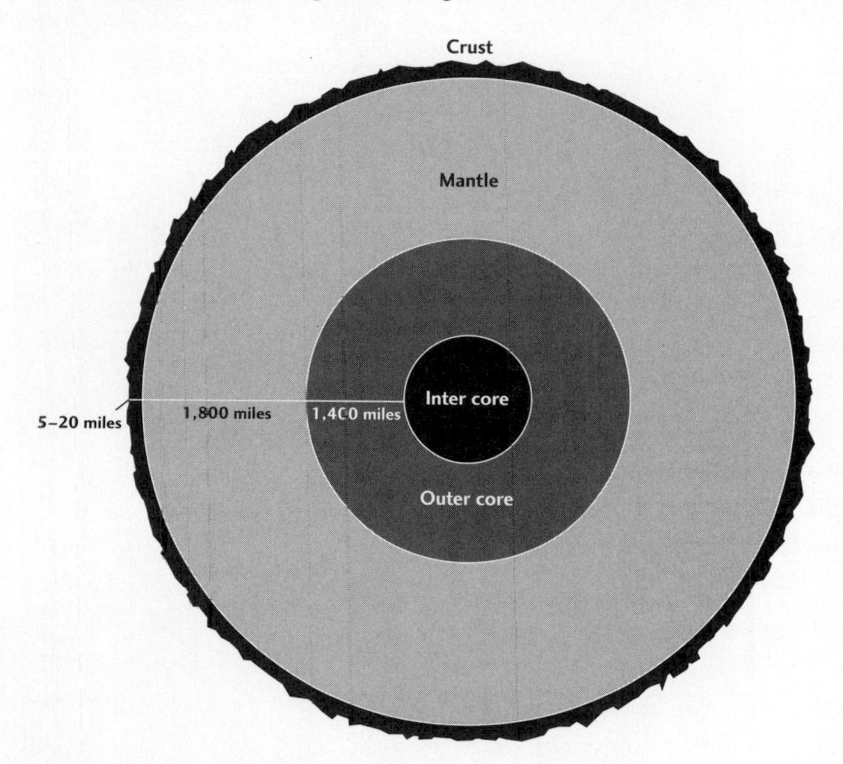

Internal Structure of Earth

has modified and moved the great landmasses, causing continents to drift and seafloors to split. To explain such a living, shifting Earth, geologists have developed a field of study based on the theory of plate tectonics. Plate tectonics suggests that much of the upper sixty miles of the Earth's surface rests on ridged plates—eight to twelve major plates and several minor ones. The plates, driven by the rising and sinking of the magma, are in constant motion. They move slowly, granted, some only an inch a year, but over time they drift apart or bump against one another, bending downward perhaps and melting into the mantle or crumpling on the edges to form mountains. Likewise, continents and ocean valleys, sitting on the huge plates, drift apart or collide or, as along the famous San Andreas Fault in California, slide past each other.

Thus, the Earth's surface, or crust, has changed. The Lubbock Lake site, for example, sits today at latitude 33 degrees 38 minutes north and longitude 101 degrees 54 minutes west. At those coordinates between 408 million and 387 million years ago—during the Devonian period—an ocean existed. What would become the modern landmass of North America lay below the equator and far to the southwest. For some 100 million years the great landmass rotated northward, moved above the equator, and, between 258 million and 248 million years ago—during the Permian period—united with other landmasses to form the supercontinent Pangaea, which reached from pole to pole. Even at that point, however, the coordinates for the Lubbock Lake site marked a spot in the ocean.

Nonetheless, the Llano Estacado's bedrock was forming. Far southwestward of the lake site's current coordinates and just above the equator, the western half of the North American landmass alternately rose up and fell back or wore down. Western mountains and basins formed as the rising and sinking of magma heaved and folded and thrust tectonic plates together and volcanoes spewed molten rock from Earth's depths. This was the Permian age.

Then, over deep time, the great Pangaea separated. The continental plates continued their slow, relentless movement across the Earth's crust, and the modern continents began to form. And, although the current Lubbock Lake site coordinates were still in an ocean, on the emerging North American continent Triassic-age rocks—called Triassic red beds—

created a thick marine rock sheet far below the Llano Estacado's present surface.

And time passed. And continents drifted. And the globe changed. And between 163 million and 144 million years ago, as the northern and southern continental landmasses moved apart, an equatorial seaway appeared. And, finally, between 84 million and 65 million years ago the continents began to assume their present shapes and current positions. At that time, global coordinates for the future Lubbock Lake site marked a spot about one thousand miles west of its present location.

The world continued to reshape itself. Magma, as we have seen, twisted and curled and moved toward the Earth's surface. Some magma simply looped back into the Earth's mantle; some solidified into the hard, solid part of the globe's crust, forming rocks.

Geologists divide rocks into three main types: igneous, sedimentary, and metamorphic. Igneous rocks formed as magma moved toward the Earth's crust. Many of those that broke the surface, the volcanic rocks, cooled quickly and thus the minerals they contained did not have time to form large crystals. Obsidian is one such rock; it is glassy but does not contain crystals. Those volcanic rocks that cooled more slowly formed rocks with tiny mineral crystals, such as basalts. Rocks from magma that did not rise all the way to the surface cooled and hardened slowly and thus formed coarse mineral grains with large crystals: granites, for example.

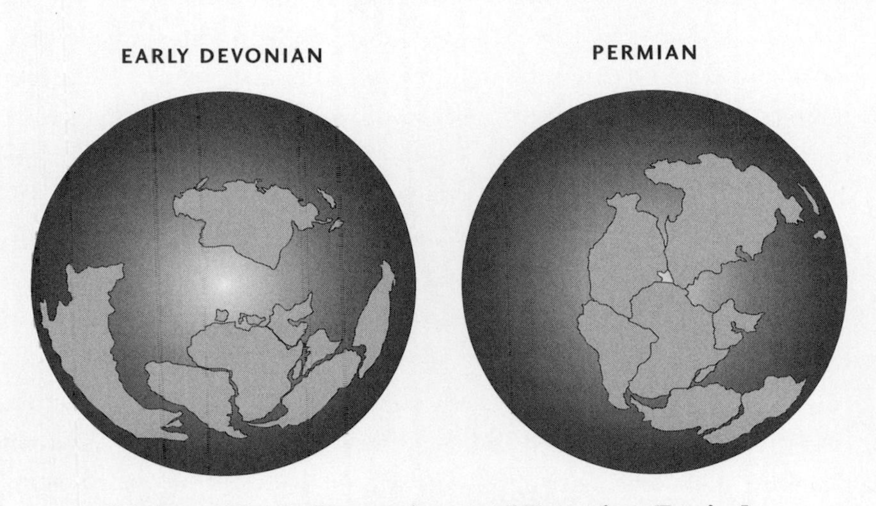

Globe at Early Devonian and Permian Periods

Sedimentary rocks consist of materials that once were part of older rocks or of plant and animal matter. Such rocks formed over millions of years as layers of loose deposits from the various materials pressed together and changed into solid rock. Sedimentary rocks formed in several ways. In one method, grains of sand from older rocks, worn away by wind or rain, became buried and under this pressure a stone-forming process eventually took place. If the pressure was severe, water might be squeezed from the buried material. Shale came from clay in such a process and siltstone from silt.

In a second method, sedimentary rocks formed from mineral deposits that had dissolved in water. The subsequent evaporation of the water caused the minerals to crystallize. The process created such rocks as gypsum, iron ore, flint, rock salt, and some limestones.

Many other limestones formed through the third method—from shells, skeletons, and other parts of animals or plants. The famous white cliffs of Dover in England came from one-celled animals that over millions of years created the chalky limestone. Coal, which developed during the same period, formed from ferns and other marsh plants that had become buried in swamps and then decayed.

Metamorphic rock is a bit different. It is rock that has changed its appearance and sometimes its mineral composition. Hot magma, pressure and heat from mountain-building, or the chemical action of liquids and gases might create such rocks. These changes occur deep in the Earth. Many rocks, including igneous and sedimentary, have gone through such metamorphism. Metamorphism, for example, can recrystallize the calcite in limestone to form marble. Similarly, soft shales and clays can harden to form slate.

At the Lubbock Lake site, sedimentary rocks dominate. But deep below the surface geologists have found both metamorphic and igneous rocks. Some date to the Permian and Triassic periods, approximately 230 million years ago. Elsewhere in the world, the oldest rocks geologists have found date to about 4 billion years ago—about the time life began.

Life began with atoms. The atoms formed bonds with one another to forge simple molecules, which first appeared in ponds and oceans. Later, one-celled organisms developed. By 3 billion years ago, in the Precambrian period, multicellular organisms had evolved, and eventually blue-green algae and other simple, plantlike organisms began to fill the oceans. The Precambrian period with its algae-filled oceans dominates 85 percent of Earth's long history.

Common Rocks

Rock	Color	Structure
IGNEOUS ROCKS		
Basalt	dark, greenish-gray to black	dense, microscopic crystals, often form columns
Gabbro	greenish-gray to black	coarse crystals
Granite	white to gray, pink to red	tightly arranged medium-to-coarse crystals
Obsidian	various colors, but black most commonly	glassy, no crystals, breaks with a shell-like fracture
Peridotite	greenish-gray	large, pipelike formations
Pumice	grayish-white	light, glassy, frothy, fine pores, floats on water
Rhyolite	gray to pink	dense, sometimes contains small crystals
Scoria	reddish-brown to black	large pores, looks like furnace slag
Syenite	gray to pink and red	coarse crystals, resembles granite but has no quartz
SEDIMENTARY ROCKS		
Breccia	gray to black, tan to red	angular pieces of rock, held together by natural cement
Clay	white, red, black, brown	fine particles, dusty when dry, muddy and sticky when wet
Coal	shiny to dull black	brittle, in seams or layers
Conglomerate	many colors	rounded pebbles or stones held together by natural cement
Flint	dark gray to buff	hard, breaks with a sharp edge
Limestone	white, gray, and buff to black and red	forms thick beds and cliffs, may contain fossils
Sandstone	white, gray, yellow, red	fine or coarse grains cemented together in beds
Shale	yellow, red, gray, green, black	dense, fine particles, soft, splits easily, smells like clay
METAMORPHIC ROCKS		
Gneiss	gray and pink to black and red	medium to coarse crystals arranged in bands
Marble	many colors, often mixed	medium to coarse crystals, may be banded
Quartzite	white, gray, pink, buff	massive, hard, often glassy
Schist	white, gray, red, green, black	flaky particles, finely banded, feels slippery, often sparkles with mica
Slate	black, red, green, purple	fine grains, dense, splits into slabs

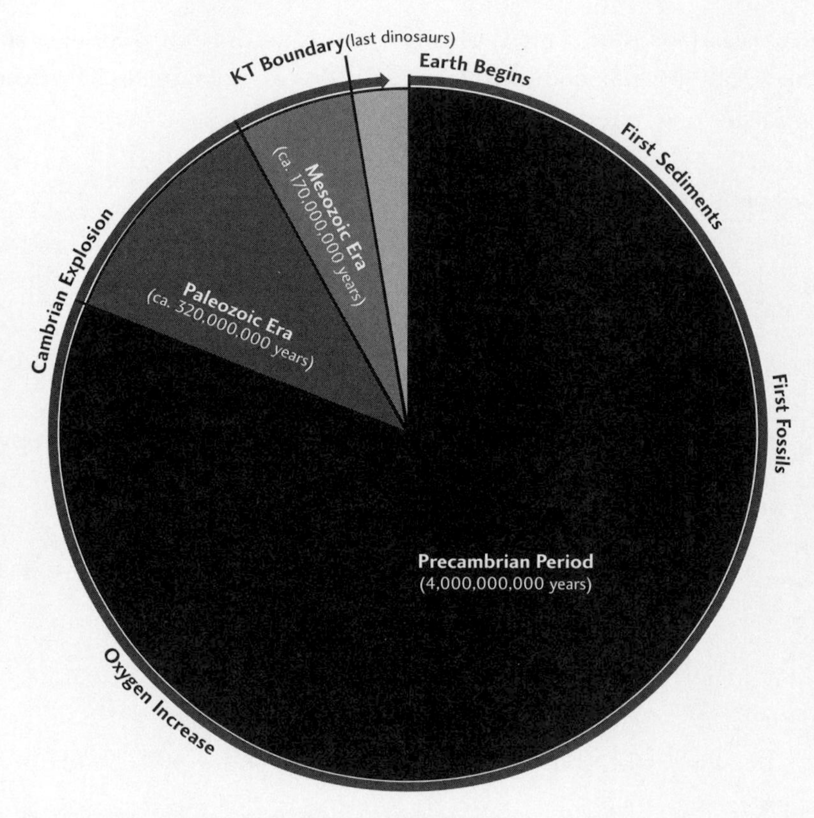

Earth History

The long monopoly did not last. Some 600 million years ago, in what is called the Cambrian explosion, a rapid proliferation of new life-forms occurred. Multiple new species followed quickly one after another. Then, about 500 million years ago, insectlike trilobites walked and hunted the ocean floor, and the first vertebrates and first fish began to appear. About 420 million years ago, primitive plants began to colonize land. Early forms of multilegged crustaceans, beetles, mites, spiders, and insects followed. Nearly 408 million years ago lobe-finned fishes first walked on land, and about 374 million years ago, the *Ichthyostega,* which combined fishlike qualities with primitive legs and other animal-like features, appeared. It was something of a transitional species between fish and amphibians.

Amphibians, able to survive both on land and in the water, dominated life on Earth for many millions of years. Most were small, such as primitive frogs. Others were huge, lumbering creatures. But whatever their size, they all remained close to their water-based habitats for a variety of reasons. Reproduction was one of the most obvious reasons—they needed the water to support the laying, incubating, and hatching of their soft eggs and to keep the eggs from drying out.

New life-forms developed. Winged insects, for example, appeared with the amphibians. Then, about 338 million years ago, during the Carboniferous period, hard-shelled eggs formed. Now, for some species at least, the need for water-based reproduction had been broken, and the earliest primitive reptiles began to evolve.

Plant life changed, too. Tall trees, particularly evergreens with outstretched branches and long, slender leaves draping downward, evolved and created thick, dark forests. On the damp forest floor grew various seed ferns and tree ferns that reached to different heights. Vines, some heavy and thick, took root as did other plant species. Here was a warm, steamy, swamplike jungle of tangled growth that admitted a few rays of sunlight and provided a rich environment for early lizards and other primitive reptiles.

But Earth was not warm and steamy everywhere. As plate tectonics transformed oceans and moved continents, weather patterns shifted, ice packs expanded and retracted, and climate and temperatures fluctuated—all over the globe. In the Earth's polar regions, for example, ice sheets have appeared, disappeared, and reappeared again throughout deep time.

Astronomical forces also influenced climate. The amount of sunlight reaching the Earth through the year, during seasonal changes, or over a decade or more had an impact on weather and temperatures. And, assuming modern weather conditions have any meaning to conditions millions of years ago, the intensity of solar radiation varied, with the poles and higher latitudes receiving less intense sunlight than regions at the equator, thus heating oceans and continents to varying degrees.

Likewise, the atmosphere—that whole mass of air surrounding the Earth—impacted climate and life-forms. Its gases, mainly nitrogen and oxygen but also the more critical carbon dioxide, circulate in complex relationships with ocean currents, mountain systems, temperature extremes, and other variables.

Again assuming that modern weather conditions have meaning for conditions millions of years ago, the major air currents moved at different latitudes in the globe's atmosphere. Today, the warmer, westward-moving trade winds blow near the equator. The cooler, eastward-moving westerlies blow between thirty degrees and sixty degrees above and below the equator. At still higher latitudes, those closer to the polar regions, the colder, westward-moving easterlies move across the globe. But the three zones are not always static, and wind currents shift, deflected by mountains or by other physical phenomena, causing a reduction or expansion of the zones. Thus, air currents, including the so-called jet streams, contribute to modern climate variations and by extension to global life-forms.

In the long view, the Earth's climate and weather have changed in numerous ways. Moreover, the transformations came with huge ramifications for plant and animal life. Across deep time, then, life-forms adapted to changing climates and weather patterns, dispersed, or died out. Thousands, perhaps millions, of separate and distinct species, such as large dinosaurs, woolly mammoths, and passenger pigeons, have disappeared from Earth. Indeed, 99 percent of all plant and animal species that once lived on Earth are now extinct. Other species, such as ginkgo trees, crocodiles, cockroaches, and sharks, have existed since long before humans, some through deep time.

But, in spite of, or as a result of, shifting climates and changing weather patterns, life-forms expanded. About 286 million years ago, at about the time of the Carboniferous-Permian boundary, along the equator where the warm trade winds blew across a wet, moist tangle of jungle ferns, primitive evergreens and other plants evolved, as did lizards and amphibians. Some of the lizards grew quite large. They also spread, although slowly, to the drier, more temperate latitudes. Dry-weather plants spread, too.

Global warming may have been a factor. The Permian period in which they lived (about 286 million to 248 million years ago) represented such a time. It saw the huge, expansive ice caps retreat toward the poles and the consistently warm and wet environments near the equator give way to distinct seasons of warm and dry climates. Away from the equator, at the higher latitudes, wet and dry periods helped to create zones where cooler temperature extremes occurred. Few large animals of the period could live at the higher, drier latitudes.

Mass Extinctions

Perhaps seventeen extinction events have occurred on Earth since life began 4 billion years ago. An event in which more than 20 percent of the marine (and a larger percentage of terrestrial) genera perished can be classified as a mass extinction. Paleontologists describe such large die-offs in different ways and by degrees of severity, such as major, intermediate, or minor.

The most extensive of the mass extinctions occurred approximately 250 million years ago at the Permian-Triassic boundary. The seas were full of fish; amphibians and reptiles dominated among the land animals; but warm-blooded, "proto-mammals" were emerging. The varied and abundant flora and fauna, both in the seas and on the land, prospered. Then, at the end of the Paleozoic era and almost all at once, a catastrophic event caused 90 percent of all species to vanish, including about 95 percent of all land animals.

The most famous extinction happened 65 million years ago at the Cretaceous-Tertiary (K-T) boundary. The most popular theory to explain it suggests that a giant meteor or comet struck the planet, triggering a subsequent crisis that ended the age of dinosaurs. In this extinction at least half (and possibly 66 percent to 80 percent) of all terrestrial and marine species died. In the oceans, for example, the very important ammonites that constitute marine plankton perished. The result: fish starved. On land many plants died. The result: the animals that fed on them starved. Many small mammals survived. Freshwater fish, turtles, and crocodiles survived as well, and snakes and lizards were almost untouched.

What accounts for the global die-off? Some scholars have suggested that, as a few of the most severe die-offs have come at regular intervals, a meteorite or similar astronomical agent in its long circular journey periodically visited Earth. Its trailing debris impacted global life-forms by striking the planet. Some people have suggested that the Earth's—and the solar system's—orbit through the Milky Way galaxy offers the best solution, for at regular times in its long journey the planet moves through various interstellar fields that affect life on Earth. Most earth scientists and paleontologists, however, suspect that global crises have occurred as weather patterns changed and climates shifted, altering the atmosphere and oceans: ice ages, for example, or volcanism, or changes caused through plate tectonics and continental drift.

Solving the mysteries of mass extinctions represents another of those wonderful puzzles for paleontologists, geologists, evolutionary biologists, and other scientists who study issues relating to deep time.

In fact, global warming at the end of the Permian period caused major animal extinctions. Many marine life-forms died out—perhaps 70 percent of the ocean species. Land animals suffered enormous losses; plant species adapted or died. The formation of the supercontinent Pangaea, mentioned earlier, further complicated the climatic trends and plant and animal events during the Permian. As Pangaea formed, ocean currents became disrupted and the cooler ocean waters could no longer aid in the dissipation of heat, thus adding to the global warming.

Years passed—millions of years. Across deep time many plants and animals in the thick, tangled, heated forests around the globe died. Layers of decomposed materials developed, and the thick deposits became the carbon-rich coal beds and valuable petroleum deposits useful to the modern world.

Also across deep time, amphibians and reptiles expanded their range of territory. They moved away from the warm, humid zones associated with the globe's lower latitudes, and lizards, particularly some of the smaller ones, ventured into the drier temperate zones associated with higher latitudes. But with potential food supplies in the drier regions less abundant, the invaders often remained small. Larger animals, on the other hand, tended to stay closer to water, where they found abundant food supplies, and as a result grew ever larger. Reptiles, especially, underwent transformations in sizes, types, and numbers.

Because they were limited to their water-based habitats, amphibians possessed few options for migration. During times of global warming, when dry seasons expanded and individual water sources disappeared, amphibians frequently died. The seasonal changes that occurred during the warming trend of the Permian period led to the extinction of many amphibian species. Plant life changed, too. While plants that were ill-equipped to withstand the changing seasons and drying climates perished, seasonal changes aided the development and expansion of other plants. New plant species developed even as others died out.

New animal species also developed—the great dinosaurs among them. Dinosaurs appeared after the global warming of the late Permian period. They existed roughly between 248 million and 65 million years ago, or during the Mesozoic era. They came to characterize much of the era's three periods: Triassic, Jurassic, and Cretaceous.

Dinosaurs represented a diverse group. They were of separate orders, genera, and species, and they enjoyed an abundance of different shapes and sizes. Moreover, through millions of years dinosaurs evolved and changed. Like plants and other animals, some dinosaur species became extinct even as others had just begun to develop.

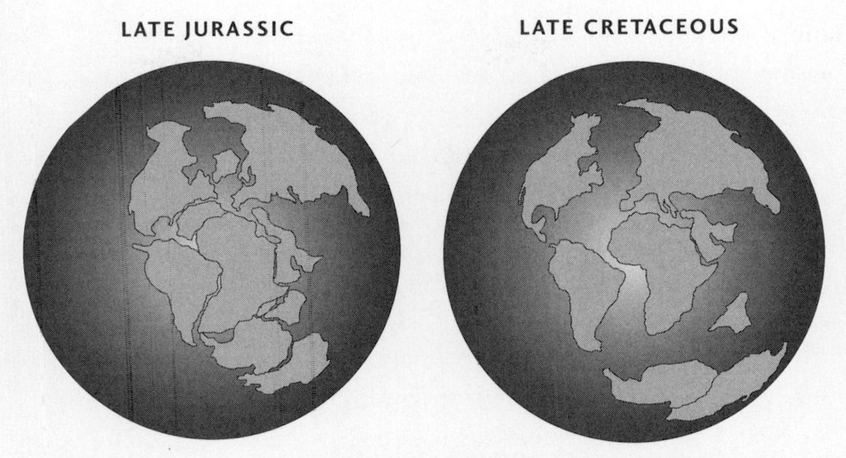

Globe at Late Jurassic and Late Cretaceous Periods

In 1841, Sir Richard Owen, an Englishman, coined the term *dinosaur.* He created it from two Greek words: *denos,* meaning "terrible," and *sauros,* meaning "lizard" or "reptile." Although originally used to describe all known dinosaurs, the word *dinosaur* today is used as a popular term for two separate biological orders of extinct reptiles: Ornithischia, an order that developed birdlike pelvic structures (bird-hipped); and Saurischia, an order that developed reptilian pelvic structures (lizard-hipped). The orders are not closely related, and they underwent distinct evolutions.

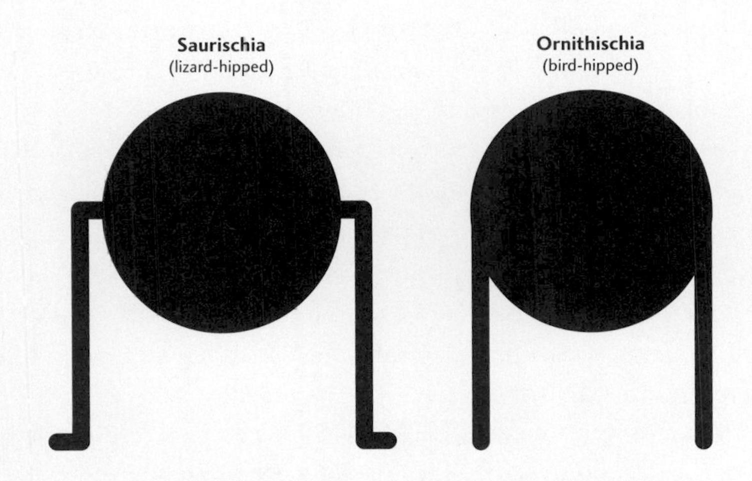

Pelvic Structure of Dinosaurs

Ornithischian dinosaurs were herbivores—plant eaters. Many of them were quadrupedal, but with their hind legs considerably larger than their front legs. The order included ceratopsians, or horned dinosaurs, such as *Triceratops;* stegtosaurs, or plated and spiked dinosaurs, such as *Stegosaurus;* and such others as duck-billed, spined, bone-headed, and beaked species.

Most saurischian dinosaurs were bipedal carnivores—meat eaters. They ranged in size from small, quick creatures, such as *Coelophysis,* a three-foot-tall Triassic dinosaur with a whiplike tail and short front limbs well adapted for grasping prey, to large and powerful hunters, such as *Deinodon,* a Cretaceous dinosaur with massive and powerful hind legs. A few saurischian dinosaurs were four-footed, plant-eating sauropods, such as *Diplodocus,* a Jurassic-aged giant that was among the largest land animals ever to have lived.

Dinosaurs did not all live at the same time. Some of the giant sauropods, such as the brontosaurus-like *Apatosaurus,* lived during the Jurassic, but by the mid-Cretaceous they had disappeared. The popular and ferocious *Tyrannosaurus rex,* on the other hand, did not even appear until the Cretaceous. But as a species, dinosaurs characterized life on Earth for 180 million years.

Dinosaurs first appeared in a world populated by two major kinds of land plants: (1) ferns and primitive evergreens and (2) seed ferns, palmlike cycads, and conifers. In some places the plants created thick, dark forests. In other places they were part of tropical swamps. Except for some of the conifers, they did not do well in drier uplands and in cooler temperature zones, places where only sparse vegetation grew.

Then, in the middle Cretaceous period, a third group of plants, flowering species, evolved. Flowering plants include grasses, fruits, vegetables, grains, roses, lilies, oak trees, apple trees, and most others that produce some sort of flower, even a tiny, colorless one. Flowering plants did not all appear at once, of course, but by the late Cretaceous they had come to dominate plant species. Consequently, food was abundant for both plant- and meat-eating dinosaurs.

Dinosaurs were not alone on the planet. As we have seen, amphibians, lizards, beetles, insects, and other animal species were present. Birds were evolving, particularly in the Jurassic period, and by the Cretaceous period birds had developed into essentially their modern form. Mammals, as hair-covered, warm-blooded, breast-feeding animals, were also present.

The first mammals appeared some 180 million years ago, about the Triassic-Jurassic boundary, but their origins may go back as far as 320 million years. They evolved slowly—very slowly—from the same ancestors that produced warm-blooded reptiles. Many early mammals were tiny rodentlike creatures, such as shrewlike animals and primitive opossums that fed on grubs, insects, and eggs. Through most of the age of dinosaurs, mammals remained small. The little and inconspicuous early mammals perhaps worried about the dinosaurs who fed on them, but in turn they may have eaten dinosaur eggs.

No dinosaurs lived on the Llano Estacado, for the high Texas tableland did not exist until more recent times—until after the end of the Cretaceous period. But well below the Llano's current surface, deep in the red bed sands and silt and mud of the Triassic period, geologists and paleontologists have discovered dinosaur remains in rocks and gravels called the Dockum Group. In the 1890s geologists dug out dinosaur bones in Dockum soils near Big Spring. Similarly, Triassic rocks with fossils of various kinds can be found along the Canadian River brakes and west of the Llano Estacado in New Mexico. Sankar Chatterjee of Texas Tech University has discovered fossils of the Dockum near modern Post, deep in the Llano Estacado's eastern escarpment. Palo Duro Canyon has also produced fossils from the Dockum Group, which represent the oldest dinosaurs in Texas.

The fossil hunters do not have a complete picture of Dockum Group dinosaurs, but they have found several varieties. One is *Chindesaurus,* an early species related to some of the oldest dinosaurs of South America. *Chindesaurus,* which Chatterjee found, was a small, long-tailed meat-eater. In Palo Duro Canyon, Edward D. Cope, a paleontologist from Pennsylvania, discovered bones and teeth of a crocodile-like reptile. Another Triassic-age dinosaur that may have lived in West Texas, one that was cannibalistic, was the lizard-hipped meat-eater *Coelophysis,* an animal that was six feet long, fast, and active. Paleontologists found fossilized remains of at least a thousand of them in New Mexico in red riverine muds that over time had turned to soft stone.

The fossil finds near Post, also located in Triassic red beds, have turned up some dinosaurs of their own. One of them is *Shuvosaurus,* apparently a lizard-hipped, bipedal, meat-eating species. The only plant-eating dinosaur of the period and place, *Technosaurus,* which Chatterjee also discovered, was a small amphibian, not as yet fully identified. Many of the bones at the Post quarry represent large-headed amphibians or reptile groups other than dinosaurs.

The fossil and geological records reveal much about the early Llano Estacado area. In the late Permian, some 250 million years ago, the surface of the region mirrored that of the modern Amazon delta. Long-vanished rivers, floodplains, swamps, and a junglelike environment intermixed with sea waters. Over millions of years the advance and retreat of the seas alternately destroyed and fostered swamps and plant life. With the most recent, southwestern retreat of the sea, the Permian period closed. No amphibian, reptile, or other vertebrate fossil remains have been found in the area, including the Lubbock Lake site, dating to the late Permian.

The Triassic period followed. By its beginning, around 230 million years ago, the sea had retreated, but the climate in the area remained warm and humid. Partly as a result, small lakes, swamps, and abundant vegetation covered a land of meandering streams in an environment that encouraged amphibian and reptilian life.

Dinosaurs from the Jurassic and Cretaceous periods are not present in the rocks below the Llano's surface. The stratigraphic record at Palo Duro Canyon, for example, represents much of the past 250 million years, but at the top of the Triassic's Dockum Group formation sits an unconformity—that is, a buried zone where either no deposition of soils occurred or erosion had removed the soils. Here on the unconformity one should find Jurassic and Cretaceous rocks, but they are not present. It is not known if such rocks were ever deposited in the Palo Duro Canyon area.

In Yellow House Canyon and at the Lubbock Lake site the stratigraphic record is similar. It shows some Cretaceous-age limestones, shales, and sandstone sitting on the Triassic red beds, but it reveals no Jurassic rocks and no dinosaur fossils. In other words, the unconformity so very pronounced at Palo Duro Canyon is widespread. In fact, no stratigraphic layer above the Triassic extends across the entire Llano Estacado.

Nonetheless, south of Yellow House Canyon a thick northwest-southeast-oriented band of Cretaceous rocks overrides Triassic material.

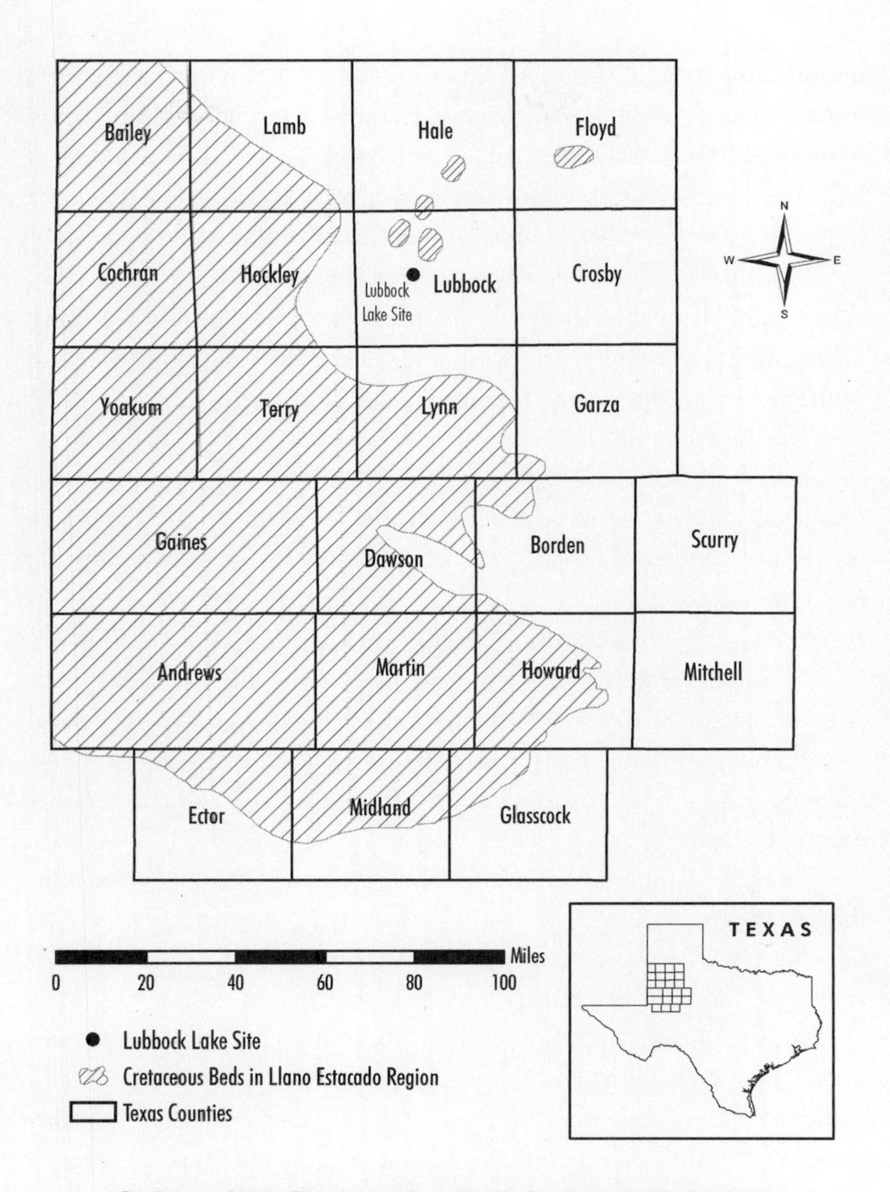

Subsurface Cretaceous Beds in Llano Estacado

Again, no Jurassic rocks are present. Here, the Cretaceous rocks and fossil record suggest that the region was a shallow sea, for fossils from the band include flat, spiral shells of various mollusks, snail-like species, and ancient squids and octopuses. Cretaceous fossils are sparse, however, and some places in the band contain no fossil records.

Consequently, while they dominated much of the world between 210 million and 65 million years ago, Jurassic and Cretaceous dinosaurs left no records of their existence at the spot marked by latitude 33 degrees 38 minutes north and longitude 101 degrees 54 minutes west—the Lubbock Lake site. Either erosion removed whatever records they had left or they were not at the place we call the Lubbock Lake Landmark.

Then they were gone. About 65 million years ago dinosaurs died out, either suddenly in the cataclysmic crash of a huge asteroid striking Earth or over time in response to declining food sources and a slowly shifting environment. Maybe a long series of volcanic explosions, such as at the Deccan Traps of India, spewed out vast amounts of lava, polluted the atmosphere, and changed the world's climate. Or perhaps other factors, such as disease or the rise of mammals, led to their demise. Trying to resolve such mysteries is what helps make history and science fascinating professions.

Whatever the cause, the mass extinctions at the Cretaceous-Tertiary (K-T) boundary created a niche for the rise of mammals and the proliferation of birds. The K-T boundary represents something of a grand frontier. Birds crossed the great divide, but their close relatives, the huge dinosaurs, did not. Mammals crossed, but many animal and plant species did not make it.

The mass extinctions of both plant and animal life marked the beginning of the modern, or Cenozoic, era with its Tertiary and Quarternary periods. The era represents the most recent 65 million years of Earth's history. If one million years is a very long time; 65 million years is deep time.

Through such deep time mammals evolved into a wide variety of animals. There came into existence, for example, the catlike Felidae family, the doglike Canidae family, and the bearlike Ursidae family. They are carnivore families. Herbivore families that evolved included such groups as the horselike Perissodactyla family and the elephantlike Proboscidea family. Like plants, mammals did not all evolve at once, and some species died out before others appeared. Still mammals moved into ecological niches that dinosaurs had once filled.

Over time mammals acquired some of the same qualities that dinosaurs had possessed. Armadillos, for instance, developed an outer covering reminiscent of such armored dinosaurs as ankylosaurs—the minivan-sized

Europlocephalus, for example. Rhinoceroses have a defensive horn suggestive of ceratopsian dinosaurs—*Triceratops*, for example. Woolly mammoths and elephants evolved into large and powerful animals reminiscent of gigantic sauropods—*Brachiosaurus*, for example. The relatively small head and very long neck of modern giraffes is also suggestive of sauropods, as some of them had long necks—*Apatosaurus*, for example. In such mammals as horses and antelopes the use of flight for defense is indicative of *Struthiomimus*, among the fastest of all dinosaurs.

As mammals evolved, the Llano Estacado's modern surface developed. The Llano's surface owes its development to the formation of the Rocky Mountains, created through the process of plate tectonics. For millions of years the North American plate, moving westward an inch or two a year, and the Pacific plate, coming from the southwest, plowed together. The slow, relentless power with which the plates heaved and smashed against one another caused the plates to wrinkle and thrust upward, thus creating mountains. Through deep time the mountains rose up, fell back, and rose again. Jackson Hole in northwestern Wyoming is a great example of a place where a mountain range fell back into the Earth's crust.

In the Rockies, individual mountain ranges are different ages. The Colorado Rockies and the San Juan range are about 5 million to 7 million years old, and they are the youngest. The Tetons are about 9 million years old. Most other ranges are much older, dating to 45 million or more years ago. Sedimentary rock comprises many ranges in the northern Rockies. In the southern Rockies many of the mountains metamorphosed into such hard and very old granites as gneiss and schist.

But, even as they were thrusting up, the Rockies began to wear away. Natural forces such as wind, rain, snow, and flood ate away the mountain landscapes, forcing eroded material in the form of sand, silt, mud, and gravel down the mountainsides. On the eastern slope the debris moved out of the mountains in streams, some of them large rivers. Especially during spring snow melts or after heavy rains, the rivers carried enormous amounts of debris and other material down the mountain slopes and out across the plains to the east.

The outwashed debris, coupled with windborne sediments, created the Ogallala formation. The Ogallala consists of such sedimentary rocks as

Eastern caprock escarpment of the Llano Estacado in Garza County, Texas, showing the caliche soils that delayed erosion. Courtesy Southwest Collection.

sandstone, limestone, clays, and silt, plus some metamorphic and igneous pebbles and cobbles that also washed down from the mountains. The huge underground formation extends from southern South Dakota in the north to the Edwards Plateau in the south (see the map on page 103). It reaches from near the foot of the Rocky Mountains eastward some three hundred miles in places. It dates from between 5 million and 10 million years ago, after some of the last of the southern Rockies, including the Sangre de Cristo range, had pushed skyward. Essentially, the Ogallala formation rests on the Triassic red beds, and it ranges in thickness from as little as a few feet in some places to seven hundred feet in others. The Llano Estacado's underground water supplies exist in the Ogallala formation.

Near the Llano Estacado's surface, the so-called caprock represents one of the more conspicuous elements of the region. Composed of sandy limestones and originating from the evaporation of calcium-rich water, the caprock remains harder than the soils above it and the sedimentary materials of the Ogallala formation below it.

Consequently, the caprock slows erosion. The Llano Estacado, as a result, has survived into modern times. But, in fact, even as it formed, the Llano Estacado was under attack. On the west, about a million years ago the Pecos River, inching northward from the Rio Grande, cut across, or beheaded, the important debris-carrying streams from the Rocky Mountains and robbed the Llano of additional water-borne deposits. Blackwater Draw is one of the beheaded fossil streams. On the east, wind and water erosion cut down the surface; reduced its elevation; produced the Rolling Plains, a region of eroded terrain with spacious prairies; and created the abrupt, dramatic, and east-facing escarpment. The erosion rate on the eastern edge is about one inch per year. On the northwestern edge of the Llano one can also find sharp uplifts, and similar, but less precipitous, escarpments of the Llano Estacado exist on its north and southwest sides.

Above the caprock are the modern soils of the Llano Estacado. The most conspicuous soils are those associated with the Blackwater Draw formation, a widespread band of soils that wind carried from the Pecos River valley and that now overlies the Ogallala formation. It varies in thickness, according to Vance T. Holliday, from a thin layer of sandy loam in the southwest to a thick layer of clay loam in the northeast. Over time, soil development coupled with deposits from various sources, such as running water, lakes, wind, and marshland, modified the Blackwater Draw formation.

In addition, here and there one can find other, less conspicuous deposits or rock formations. They interrupt the Blackwater Draw and include the Blanco and Tule formations. The Blanco, which is the most

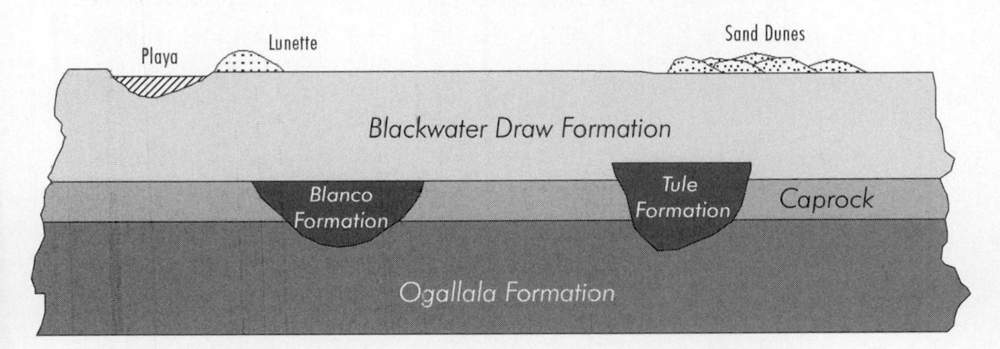

Representative Stratigraphic Cross Section of the Llano Estacado Showing Some Geomorphic Relationships

pronounced and sometimes extends down through the caprock and into the Ogallala formation, dates between 5 million and 1.6 million years ago. The other formations are more recent, dating from 1.6 million years ago. Deposits of various kinds, but especially lake and wind-borne ones, created such formations. Playa basins, or wet-weather lakes, also cut into the modern surface of the Llano Estacado, and sand dunes exist on it.

The sand dunes fall into one of two basic types. One of them, the lunettes, is associated with playa basins (shallow, wet-weather lakes). Found on the east, south, and southeast sides of the basins—the leeward side—the lunettes formed as sediments from sand, dirt, and mud blew in from the playas. The large dune fields, such as the Mescalero, Muleshoe, and Monahans, consist of sands that originated in the Blackwater Draw formation and date to about ten thousand years ago.

Rivers cut deep canyons into the eastern edge of the Llano Estacado. Some of the major cuts include Palo Duro, Tule, Quitaque, Blanco, and Yellow House canyons. Blanco Canyon with its Running Water Draw and Yellow House Canyon with its Blackwater and Yellow House draws form significant portions of the upper Brazos River system and stand, as does the Lubbock Lake site, in the very heart of the Llano Estacado.

In terms of deep time, the Llano Estacado is very young. It formed long after the major dinosaurs had disappeared. Debris from the wearing down of the Rocky Mountains created it. Additional wind- and water-borne deposits altered it. The Pecos River cut its western edge in a relatively straight line, but river drainage on the eastern edge created a series of large canyons that thrust far into its heartland. The Canadian River represents its northern edge. On its southern edge the Llano Estacado contains no significant topographical boundary, but here the high landform merges into other provinces, particularly the Edwards Plateau.

Until the 1930s, grass dominated the Llano Estacado's plant life. In fact, grass cover has characterized the region since the Llano's creation some 5 million years ago. The grasses have changed over time, of course, with dominant species moving and shifting as climate and temperatures changed in response to altering weather patterns.

Much the same is true of other plant life. Trees, such as pinyon and ponderosa pines and scrub oak and mesquite, in response to changing climates and weather patterns, have advanced and retreated in their range of territory. Cottonwoods and hackberrys, which today are common in the Brazos River canyons, have not always been part of the Llano Estacado or Lubbock Lake landscape. Cactus plants and prairie flowers have a similar history.

Glaciers, their advance and retreat, are partly responsible for the changes in landscape. Modern-era glaciers, which form and move in response to environmental factors, date to the K-T boundary about 65 million years ago when the Earth's temperatures began—very slowly—to cool. Many periods of glaciation occurred, and the whole process took time. By 1.6 million years ago, at the beginning of the Pleistocene epoch, Earth had become very cold. Ice sheets covered much of the northern part of the globe and high mountains everywhere.

Then, one after another, four great glacier systems and many smaller ones pushed southward through North America and northern Europe. They began about 600,000 years ago with one of them—the second, or Kansan, as we call it in North America—lasting some 156,000 years. Between the long cold periods, the glaciers retreated and tens of thousands of years of global warming occurred. The last of the four great glaciers, the Wisconsin, extended from about 70,000 years ago to about 10,000 years ago. As none of them extended below modern South Dakota, however, no glaciers reached the Texas High Plains and the Lubbock Lake site.

At their peak, the glaciers turned enough water into ice that ocean levels dropped—sometimes three hundred feet. The lower sea level created a wide stretch of land, called Beringia, in the shallow Bering Sea area between Asia and North America. Here, before melting glaciers flooded the land again, ancient horses and camels, which had originated in North America, crossed into Asia. Early species of elephants, bison, deer, and bears, which had evolved in Europe or Asia, in turn migrated to North America.

As the glaciers pushed south, they forced animals to move as well. Accordingly, such species as llamas, giant ground sloths, and armadillos migrated to South America, and cold weather mammals such as woolly mammoths and arctic musk oxen moved from Canada into the Americas.

During the warm, interglacial periods, animals followed the melting ice northward.

Animals of the Pleistocene set the stage for human occupation of the western hemisphere. Hunters, foragers, and fishermen followed the large mammals across the coarse northern grasslands or used small boats to skirt along icy shores. Near the end of the last glaciations, humans entered North America.

★

Chapter 2

The Paleoindian Period

Hunting Big Game

ARCHAEOLOGISTS and others usually describe the Paleoindian period as extending from the first human occupation of the western hemisphere to the beginning of the Archaic period, which at the Lubbock Lake site was about 8,500 years ago. They do not agree—they do not seem to be even close to agreeing—on just when and where humans first arrived in North America, how rapidly the Paleoindians moved through the continent, or how far and wide the early cultures spread. They agree only that Paleoindian people were hunters and foragers, and perhaps they do not always agree on that.

Humans have always been hunters. Such early humans as *Homo erectus, Homo neandertalsis* (Neanderthal Man), and *Homo sapiens* (Cro-Magnon Man) all pursued game animals. If their cave paintings and pictographs mean something, they may have preferred to hunt large animals. Modern-day sport hunters, who, like their ancestral counterparts, may take squirrels and rabbits, generally prefer to hunt deer, elk, bear, and other big game. Little has changed in that regard. What has changed are the weapons, tools, and methods of hunting. Lifeways, worldviews, cosmologies, and life patterns have also changed, and, of course, the game animals have changed.

Neanderthal people lived from two hundred thousand to thirty thousand years ago, during the middle Pleistocene epoch. They evolved in Europe and spread to the Near East and to Southeast Asia before disappearing when the last glacial advance reached its peak. In Europe they lived in an environment dictated by the advance and retreat of the huge glaciers. At times, as when glaciers were in retreat, warm temperatures, such as today, encouraged the spread of forests and caused large animals to disperse. At other times, as when glaciers advanced, very cold temperatures wrecked forests but encouraged the formation of grassy, tundralike plains that supported large, plant-eating animals. Neanderthal people manufactured stone and wood tools, built fires, and buried their dead, sometimes with flowers. They hunted and gathered whatever food supplies they encountered in their environment.

Cro-Magnon people and early Asian hunters of the late Pleistocene, thirty thousand to ten thousand years ago, were different. They made use of a sophisticated tool kit. It included awls, hide scrapers, knifelike blades with points, leaflike blades, and other items made from stone, antler, bone, and wood. Because they paid close attention to seasonal changes and the varying plant resources, the hunters moved to take advantage of foods found in different environments. Within their range of territory they followed and hunted large animals. They manufactured their clothing and other items from the skins of the animals.

Some of these people moved into modern Siberia with its steppe tundra, pursuing woolly mammoths, woolly rhinoceroses, ancient bison, horses, reindeer, and other grazing animals. Some of them were probably foragers of small game, fish, fruits, and nuts. Collectively they represent some of the ancestors of the Paleoindians, who were the first humans in the western hemisphere.

The first humans in the Americas, however, remain an illusive bunch. They may have been diverse groups of various ethnic and racial backgrounds. Archaeological discoveries over the past two decades have revealed enough information to suggest that the earliest humans arrived at

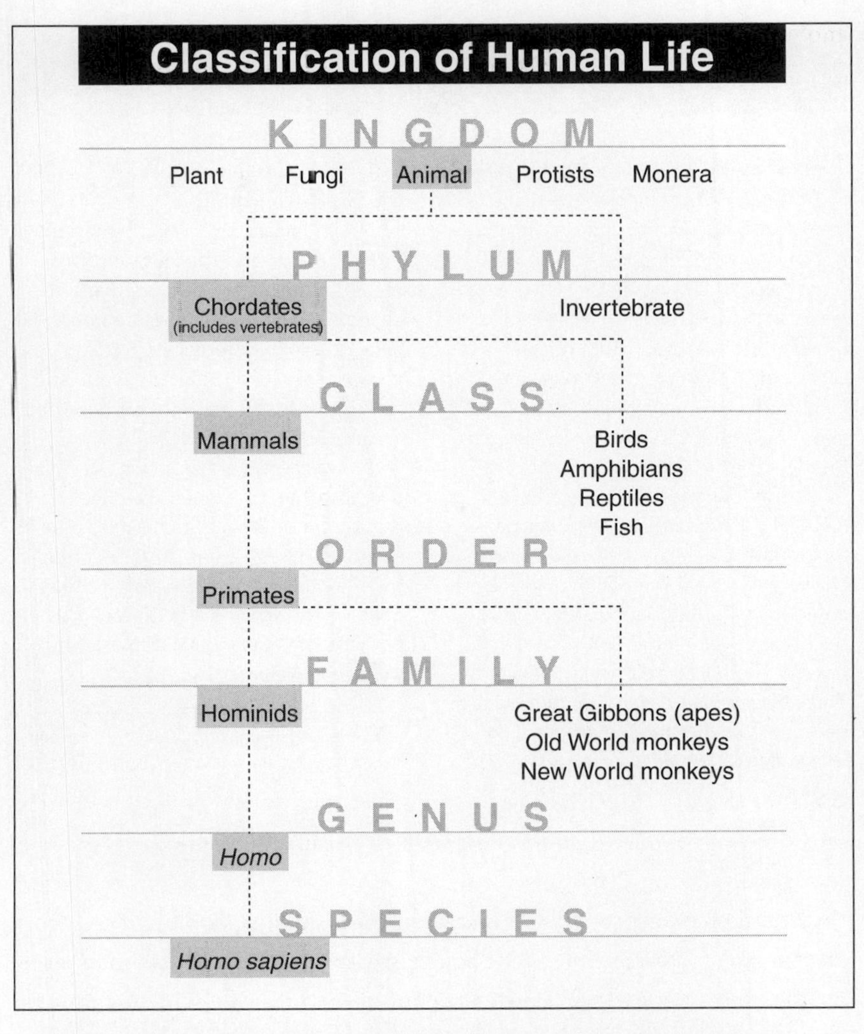

The other kingdoms are plant, fungi, protists (bacteria, protozoans, some algae), and monera (all prokaryotes). Source: Vincent Courtillot, *Evolutionary Catastrophes: The Science of Mass Extinction* (Cambridge: Cambridge University Press, 1999), 1, n2.

different times and in different places. There are recurring claims as to who came first, claims for Africans, for Australians, for Asians of non-Mongoloid stock, and for Ice Age Europeans—the Solutrian culture—from what today are France and Spain. Many, but not all, of the claims are weak on both geographical and chronological grounds.

Mitochondrial DNA and Migration to the Americas

For several years scholars have been using DNA, particularly mitochondrial DNA, to find clues for human migration, including passage to the Americas. It is a strategy that combines genes with fossils. Results from the combined strategy suggest that humans came out of Africa about two hundred thousand years ago, spread to Europe and Asia, and eventually moved through Beringia to the Americas.

The mitochondrial DNA (mtDNA) method is tricky and perhaps not perfect, but it reinforces the fossil record. Unlike ordinary nuclear DNA, which is mixed and shuffled during sexual reproduction, mtDNA—the "Eve Gene"—is passed whole from mother to child. Thus, the mtDNA of modern people can be traced back through matrilineal lines to a common female ancestor who lived in Africa. Males are mostly invisible, for with rare exception mtDNA reveals only the history of females.

An important key is that mutations in mtDNA occur, and the mutations form recognizable patterns. By studying the changing patterns, scholars can determine how long different population groups—such as Yupics in Siberia and Aborigines in Australia—have been separated.

Now scholars are examining fossilized bones and teeth—when they find them—for mtDNA evidence. Most such evidence is fragmentary and degraded, but if archaeologists can match the evidence with traditional fossil records, perhaps the human migration patterns will become clearer still.

Source: Joel Achenbach, "Who Knew?" *National Geographic* 204, no. 6 (December 2003): 1.

The best evidence still suggests that Paleoindian people crossed from eastern Asia to North America. They may have traveled in small boats along the icy coast or come over Beringia, a one-thousand-mile-wide—about the distance from New York City to Omaha—stretch of land connecting modern Siberia and Alaska. They probably came in small, scattered bands of interrelated families, fishing and hunting and using boats to traverse open seas between offshore islands. Or they may have slowly hunted and foraged their way across Beringia at a time when it represented a place of climatic extremes with long, fierce winters and short, warm summers.

A harsh land, Beringia contained a steppe-tundra characterized by marshy lowlands. Warm Pacific currents kept it free of glacial ice. It contained many lakes, marshy bogs, cold-water swamps, and rivers. On it grew coarse grasses, sedges, mosses, and other plants that provided an almost continuous ground cover and supported the plant-eating animals that Paleoindian people hunted. During its peak existence, Beringia may have included some woodland, perhaps cottonwood scrubs and birches and possibly some conifers later. Trees may have been necessary for fires to warm temporary homes, but people may also have used animal dung for cooking and heating. Beringia's southern shores may have held a few scattered settlements whose inhabitants hunted sea mammals and fished, but only a few humans probably lived there during the peak years of the last, or Wisconsin, glaciation. Still, its grasslands supported a wide range of arctic animals, including bison, woolly mammoths, caribou, reindeer, and musk oxen. Horses and camels, making their way to Asia, were there as well. Predators, such as ancient wolves and saber-tooth tigers, fed on the plant-eaters.

Humans hunted their way across Beringia. With animals scattered over the land but usually close to rivers and lakes, the Paleoindians moved eastward. Some people hunted along favorite lakes or streams repeatedly before pushing on again. Some followed the southern beaches where seals, sea otters, and other water-based mammals were abundant. Some groups did both. Others crossed the shallow sea in boats, moving from island to island through the Kuriles of Asia and the Aleutians of Alaska. In such fashions—in slow stages—Paleoindian people made their way along the coast or across Beringia to higher ground in North America.

In North America during the late Pleistocene, or Ice Age, period, much of Alaska and parts of western Yukon Territory, which is the Alaskan Refuge, was ice-free. Glacial ice existed in central Canada, called the Laurentide ice sheet, and high up in the coastal mountains, called the Cordilleran ice sheet, but an ice-free area between the two massive ice sheets extended from Alaska through the Yukon to below the glaciers. A thin strip along the Pacific Coast may also have been free of ice.

Humans trailed game animals southward. Some people, as indicated above, used small boats to navigate the shoreline and to move from one offshore island to another. In this way they may have traveled quickly toward South America.

Perhaps most people moved through Alaska rather than along the coast. Through several generations they could have hunted and foraged their way toward an ice-free passageway located in the region just east of the Rocky Mountains. With rising temperatures fifteen thousand years ago, Laurentide ice melted rapidly and a corridor may have widened, making life and travel through it possible, if not particularly attractive. Still, game animals moving south entered a passageway and people followed.

Just when the southern movement occurred remains a matter of debate. Some experts suggest very early dates: fifteen thousand to thirty thousand or more years ago. Most argue for a more recent period, one closer to twelve thousand or thirteen thousand years ago. Using dental information, some scientists have postulated three separate periods or waves of migration: more than thirty thousand years ago, about twenty thousand years ago, and twelve thousand years ago.

In any case, over time some humans pushed below the Laurentide and Cordilleran ice sheets onto the Alberta, Saskatchewan, and Montana plains. Then, about eleven thousand years ago temperatures in North America dropped again in what some call the Younger Dryas, a near-glacial cold snap of a thousand years. It did not close off the corridor, but its conditions, which may have brought wetter circumstances to the Texas High Plains, allowed occupation of the Great Plains by large animals and the people who pursued them.

Thus, once below the glaciers the Paleoindians entered an attractive environment. They found extensive steppe-plains with a generally uniform climate with relatively mild temperatures, moderate rainfall, and no seasonal extremes. Such conditions are ideal for the creation of premium grassland-grazer habitat. On the Great Plains there was tall, coarse grass with scattered trees and some thick forests. Hunters found lakes, swamps, marshy grass-lined sloughs, and beaver meadows dotting the wet, fertile landscape. Far more humid than now, the western ranges contained at that time a rich, coarse grassland environment of freshwater ponds, lakes, and streams.

Across the verdant range, Paleoindian people hunted a variety of animals. In the far north the long-haired woolly mammoth remained among their favorite prey. Farther south they chased other species of mammoth, particularly the Columbian mammoth. On the eastern edge of the Great Plains, they sought mastodons—the stocky, straight-tusked, elephant-like, forest-dwelling animals that had become extinct in Europe and Asia. They also pursued the huge, elephant-sized but slow and non-carnivorous

Short-faced bear as depicted in the outdoor, life-size sculpture at the Lubbock Lake Landmark. Photo by Tony Gleaton. Courtesy Southwest Collection.

ground sloths; massive, straight-horned bison; musk oxen; camels; horses no larger than llamas; caribou; giant beavers; and several kinds of bears. Saber-tooth tigers and other large cats and dire wolves hunted the same game animals. In addition, the people found modern animals such as big-horn sheep, elk, foxes, ground squirrels, prairie dogs, pronghorns, and birds—some of which were larger than eagles.

The people moved across the plains on foot. They probably fashioned wooden and bone clubs and long spears. They continued to seek large animals and continued to use the animals' flesh for food, the hides for clothing and other purposes, and the bones for tools and weapons. On the Texas High Plains, their primary targets remained mammoths but, as indicated above, included now extinct species of horses, camels, ancient bison, giant sloths, tapirs, and other game. And, of course, they hunted smaller animals, such as jackrabbits, which may have been an important part of their commissary.

Paleoindian people spread over the continent, entering Mexico or pushing toward South America. Some probably arrived by boat having

Columbian mammoth as depicted in the outdoor, life-size sculpture at the Lubbock Lake Landmark. Photo by Tony Gleaton. Courtesy Southwest Collection.

traversed the Pacific shoreline. A site in Monte Verde, Chile, along a low-lying creek reveals that the people who lived there, probably as early as 12,500 years ago, hunted game animals and gathered vegetable foods. They lived in houses and used wooden and stone tools and weapons. The site represents one of the earliest settlements in the New World.

In North America one of the early settlements occurred in western Pennsylvania, a site called the Meadowcroft Rockshelter. Dated to perhaps 14,250 years ago, it was occupied for thousands of years. The Paleoindian people of Meadowcroft Rockshelter hunted and gathered vegetables, used wooden and stone tools and weapons, and made projectile points with a small, lance-shaped design.

Over time Paleoindian traditions evolved. Most groups of people remained, or became, foragers of small game, fish, fruits, and nuts. Some remained, or became, big-game hunters. In either case, their descendants moved into diverse ecological zones and spread throughout the Americas. People adjusted to local conditions and made changes, some almost imperceptible, in their habits and life patterns. Their nonmaterial culture, probably including music, religion, and folklore, also changed.

Pre-Clovis Cultures

The older, predominate theory for people entering the western hemisphere has humans crossing from Siberia to Alaska as big-game hunters about twelve thousand years ago. The people crossed the Bering Strait beginning when it was a one-thousand-mile-wide stretch of land, made their way through Alaska, and then migrated south through an ice-free corridor to the High Plains of North America, using fluted, stone projectile points to hunt large game animals. They spread across North America, pushed south through the highlands of Central America, and reached the southern tip of South America at least ten thousand years ago.

Over the last few decades or so, some archaeologists have challenged the prevailing view. A few of them, such as Alan Bryan and Ruth Gruhn, claim that people using spears without projectile points appeared in the Americas more than thirty thousand years ago. The Monte Verde site in Chile and the Meadowcroft Rockshelter site in Pennsylvania, some scholars claim, reveal evidence for pre-Clovis humans in the Americas, with the Monte Verde site occupied perhaps sixteen thousand or more years ago. Moreover, many scholars now insist that humans came to the Americas in boats, using the vessels to sail through the Kurile and Aleutian islands to Alaska and then south along the Pacific Coast.

Were there pre-Clovis groups in the western hemisphere? "Probably" is the safe answer. Anna Curtenius Roosevelt of the University of Illinois, Chicago, for instance, believes that many archaeological sites show valid evidence for cultures contemporaneous with, but different from, Clovis. The sites include Broken Mammoth and Walker Road in Alaska, Quebrada Jaguay in Chile, and Caverna da Pedra Pintada in Brazil. But nearly all purported pre-Clovis sites, including Putu, Old Crow, and Bluefish Caves in Alaska (all good candidates for pre-Clovis), have flaws. Some, such as Putu and Bluefish Caves, contain little evidence of human presence; some, such as Meadowcroft Rockshelter in Pennsylvania, have produced questionable or inconsistent dates; some, such as Monte Verde in Chile, have other problems that cast doubt on their validity.

Some archaeologists who argue for a pre-Clovis colonization accept the time of arrival as a bit prior to twelve thousand years ago, but they challenge the big-game hunter thesis. These scholars, including Professor Roosevelt, argue that the first humans entering North America were foragers of small game, fish, berries, and nuts rather than specialized big-game hunters. Such hunter-foragers manufactured triangular-shaped projectile points that could be used as knives as well as spear points. Their descendents, the argument goes, reached to all parts of the Americas. The Clovis was only one of the descendant cultures.

Obviously, the controversy over when and how humans colonized the Americas has not been resolved, but many archaeologists now reject the Clovis-first model.

Sources: Robson Bonnichsen and Karen L. Turnmire, eds., *Ice Age People of North America: Environments, Origins, and Adaptations* (Corvallis, OR: Center for the Study of the First Americans, 1999), 1–21; Anna Curtenius Roosevelt, "Who's on First?" *Natural History* 109 (July–August 2000): 76–79; David Hurst Thomas, *Skull Wars: Kennewick Man, Archaeology, and the Battle for Native American Identity* (New York: Basic Books, 2000), 167–74; Bradley T. Lepper and Robson Bonnichsen, eds., *New Perspectives on the First Americans* (College Station, TX: Center for the Study of the First Americans, 2004), 1–3.

On the Texas High Plains, including the Lubbock Lake site, at least three distinct cultural traditions emerged: Clovis, Folsom, and Late Paleo-indian, sometimes called Plainview or Firstview. Dating each of them precisely remains difficult, but the dates presented here, 11,500 years ago to 8,500 years ago, follow the findings of Vance T. Holliday and Eileen Johnson, who have worked extensively at the Lubbock Lake site and across the Llano Estacado.

The Clovis tradition, which spread through much of North America, possibly came first. Archaeological investigations over the past two decades, however, are challenging the Clovis claim to priority. Nonetheless, after about a thousand years of people having arrived below the glaciers, small groups of Clovis people lived at sites across the continent. They hunted various game, large and small, and gathered vegetable foods. If they lived along coastal plains, they may have fished. They camped along rivers, streams, lakes, and other water sources where game animals came to drink. In seeking food sources, they ranged over very large territories. They stopped at caves and rock shelters, places where they may have spent the colder winter months. They also buried their dead here and created sacred rock art.

On the Llano Estacado, the Clovis tradition lasted about 600 years, from roughly 11,400 years ago to about 10,800 years ago. Evidence for the Clovis culture exists at the Lubbock Lake site, Blackwater Draw in eastern New Mexico, and several other places.

The Clovis tradition is named for an archaeological site near Clovis, New Mexico. In the summer of 1932, A. W. Anderson and George O. Roberts, area residents, discovered several fossil bones and artifacts in Blackwater Draw, a part of the upper Brazos River drainage. The men carried their finds to Edgar B. Howard, an archaeologist from the University of Pennsylvania who was working a cave site west of Carlsbad in southeastern New Mexico. Howard, before heading east at the end of the summer field season, stopped at Blackwater Draw. Upon viewing the site, he determined to come back in the spring. But in the fall a road construction company, digging for gravel in the draw, found additional fossilized animal bones. Thereupon, Howard returned in November to inspect the new finds. Several months later, in the summer of 1933, Howard and several other scientists began systematic excavations in Blackwater Draw. They found, among much other evidence, several man-made projectile points embedded in animal remains. In effect, they had discovered the Clovis culture.

Projectile points characterize the Clovis culture. Expertly made, three to four inches long, and sharp-edged, the beautiful but deadly flint points often contained a short groove—a flute—along the lower face and a small notch at the base. The groove may have been designed to help hold the point to the shaft or fore-shaft. Hunters reused the distinctive points, and stone workers resharpened the points as well as their knives by flaking off tiny bits of rock.

Stone workers preferred various fine-grained rocks for their tools and weapons. Clovis-tradition groups on the Llano Estacado favored Alibates flint, an agatized dolomite, from the Canadian River area northeast of modern Amarillo. Sometimes an entire band might travel to the Alibates quarry. At other times projectile-point and knife makers acquired favorite rocks from other local sources or through trade, some of it over long distances. Tecovas Jasper, sometimes called Quitaque flint, was a popular stone resource. Other fine-grained rocks, such as chalcedony and obsidian, existed at places far from the Southern High Plains.

Clovis people were effective hunters and foragers. They took a wide range of animals and collected wild plant foods. On the Llano Estacado they often hunted smooth-skinned mammoths and bears, but deer, ancient horses and camels, sloths, pronghorns, small rabbits, and other animals also formed part of their diet. When available, the ancient, straight-horned bison (*Bison antiquus*), which were about 30 percent larger than modern species, represented the mainstay of their economy, providing food, clothing, and a source for tools. They collected plant foods along the side draws, swamps, and bogs that characterized the upper Brazos River's Blackwater, Yellow House, and Running Water draws.

But mammoths must have been a favorite prey. Like modern-day elephants, they moved often, traveled in small matriarchal herds, and returned year after year along familiar trails to favorite water holes, food sources, and salt licks. By observing, following, and studying the animals, Clovis people might divide one from the herd, catch one off-guard, or trap

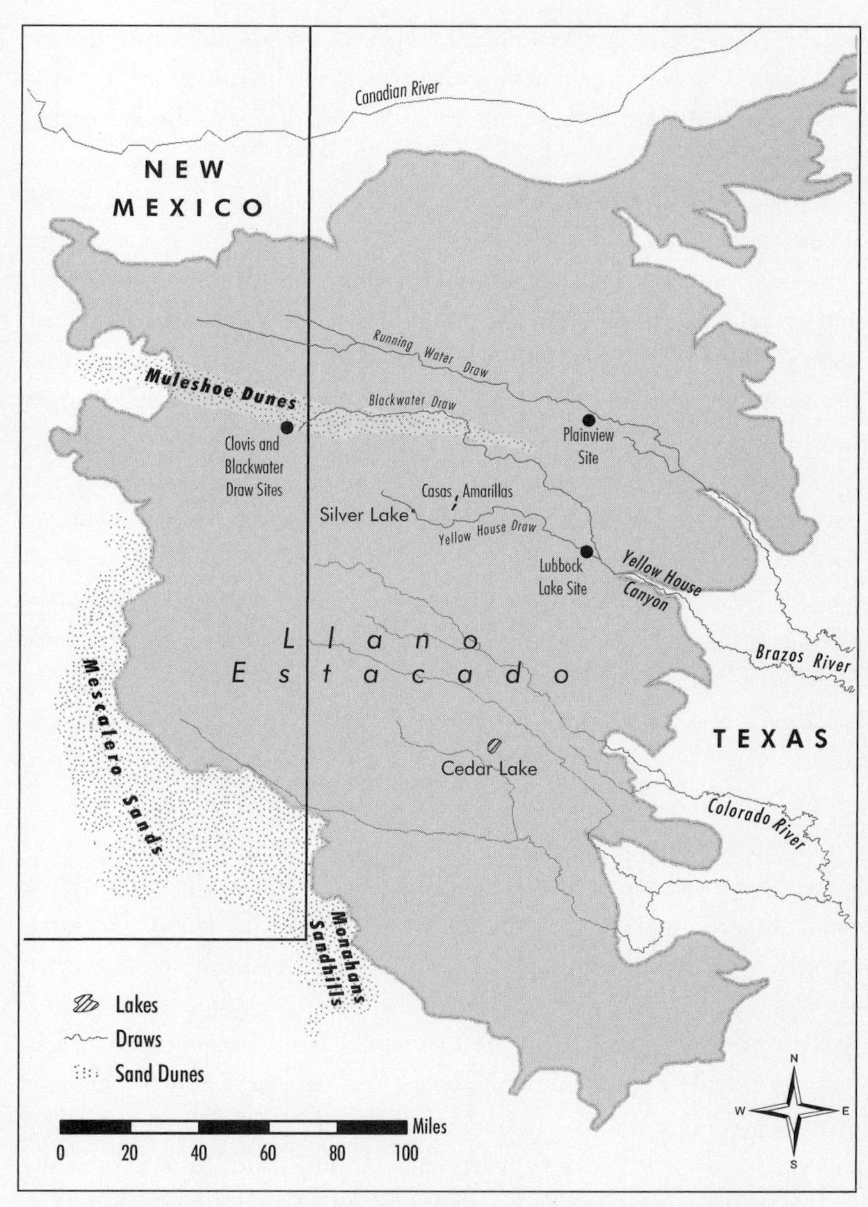

Llano Estacado Showing Upper Brazos River Region

one in a boggy spot of ground. A downed mammoth provided a lot of meat. Some of it the hunters ate immediately, some they dried, some they carried off, and some, at least in winter months, they cached. Fat provided cooking oil. The bones and tusks became tools and weapons. Skins became clothing. A large mammoth kill might provide sustenance for several weeks.

Favorite weapons must have included spears and lances. As a way to improve their hunting, Clovis people adopted the *atlatl,* a spear thrower that hunting societies had used for thousands of years. A simple but clever instrument, the atlatl was a short stick with a hook or notch on one end against which the spear was set. A hunter held the spear with his thumb and forefinger and the handle of the atlatl with his other fingers. Then, with an overhand motion and a snap of the wrist the hunter dispatched the weapon toward the target. In effect an atlatl increased the length of one's arm, and, thus, increased the spear's velocity, thrust, and accuracy.

Hunters using atlatls and Clovis points most likely lived in scattered bands. Probably thirty to fifty persons from several interrelated families composed each band. The people maintained small, informal sociopolitical organizations, for larger ones were not needed. They moved often, following the highly mobile big-game animals along their familiar haunts. Other than their tools and weapons, they carried few possessions.

Nuclear families would have been small, probably not more than the parents and two or three younger, unmarried children. Men spent much of their time pursuing large game animals or such smaller ones as rabbits, muskrats, and coyotes. They manufactured tools and weapons. Women and children collected seeds and nuts and dug roots of various plants. They also assisted in the chore of butchering and processing animal carcasses. They prepared the meals and took charge of the dwellings and the limited household possessions. During summer months some of the scattered bands must have come together for social, political, and other purposes—such as finding or arranging marriage partners or conducting large communal hunts.

The physical stature of Clovis-tradition people is unknown. There are few human skeletons of the period from which to make an assessment. One might speculate that from a modern viewpoint the people were short, stocky, and powerfully built. Men rarely stood over five feet, seven inches tall. They possessed thick, muscular thighs and broad shoulders. Women were shorter, but also enjoyed strong, round physical features. Skin color, influenced by outdoor living, probably tended toward reddish chocolate. If presuming they had a Siberian background, the Clovis people may have possessed many Asian characteristics as well, such as straight black hair.

Population numbers are also speculative. Both pre-Clovis and Clovis groups probably saw their populations increase rapidly. And, as a result, they filled the continent rather quickly. Nonetheless, with people scattered over a wide area, the population density must have remained small.

People of the Clovis tradition visited the Lubbock Lake and Blackwater Draw sites to camp and hunt. They returned often to both places. They did not maintain permanent camps at either of the popular hunting spots, but they killed and butchered animals at the wet, swampy sites. Sometimes they may have stayed a month or so, but that is uncertain. Neither place represented a massive kill site, such as at Heads-Smashed-In, Alberta, where people repeatedly drove large numbers of bison over a high cliff.

Rather, at the Lubbock Lake site Clovis hunters, with some exceptions, took single animals. Perhaps four or five men and older boys worked together in the hunt, or perhaps two men, with one of them distracting an animal, such as a deer or elk, and the other moving in for the kill, teamed up to dispatch the prey they had been carefully stalking. Sometimes women assisted.

Clovis people were at the Lubbock Lake site at least by 11,100 years ago, and perhaps earlier. Archaeologists found mammoth bones showing cut lines and tool use, but no lithic tools, corresponding to the date. They found a Clovis projectile point near a mammoth bone bed, a point that Clovis hunters had resharpened to use as a butchering device.

In addition to mammoths and bison, other animals used the Lubbock Lake site during the Clovis period. Archaeologists have found evidence of blacktail jackrabbits, blacktail prairie dogs, snapping turtles, yellow mud turtles, Canada geese and snow geese, teal and pintail ducks, turkeys, plains pocket gophers, muskrat, gray wolves, and many others, as well as now extinct giant armadillos, short-faced bears, short-legged horses, and camels. Clovis people probably made use of all such animals in their menus.

Clovis-era people lived in all parts of North America until about 10,800 years ago. Then, rather abruptly, they, or at least the features that identify them, faded. Most likely, as they adapted to new conditions, their

The Blackwater Draw Sites

The **Blackwater Draw** archaeological sites are located near Portales, New Mexico, on the extreme eastern edge of the state. A broad shallow valley without through-flowing water, the draw extends southeastward through the Llano Estacado and joins Yellow House Draw at Mackenzie Park in Lubbock, Texas, to form Yellow House Canyon.

The Blackwater Draw sites in New Mexico exist over a twelve-mile stretch where the draw runs through the western Muleshoe sand dunes in upper Roosevelt County. Archaeological investigations have occurred in several places in the dunes and in the valley walls and floor. The three major sites include Anderson Basin #1 and #2 and the Clovis site (or Blackwater Draw Locality 1).

The Clovis site is located in a gravel pit on the draw's north rim where Blackwater Draw is two miles wide and about forty feet deep. A shallow ravine leads southward from the pit (an ancient basin) to the main draw. Discovered in 1933, the site is significant for studies of the Paleoindian period and gives its name to the Clovis culture. From it, mammoth and bison bones, Clovis and Folsom artifacts, and other archaeological materials have been recovered and studied. Unfortunately, mining for gravel has destroyed much of the site.

Anderson Basin refers to a stretch of Blackwater Draw about six to eight miles downstream from the Clovis site. There, blowouts in the valley floor had exposed old bones and elephant-like tusks. Subsequent archaeological investigations revealed fossils of late Pleistocene megafauna, especially mammoths. Little or no evidence suggests that human activity at Anderson Basin was associated with the ancient animals.

Eastern New Mexico University in Portales now protects the sites and controls much of the fieldwork and excavations. Gravel mining occurs at the Blackwater Draw sites, but archaeological investigations have continued on a regular basis since the Clovis discovery in 1933.

---

ways of life evolved and their societies changed. Nonetheless, reasons for the disappearance are a matter of wonderful controversy. Some scholars maintain that as efficient big-game hunters Clovis people killed off the large animals, a development that, in turn, led to a sudden decline in the Clovis population.

Others are not so sure. About 10,800 years ago, they suggest, mass extinctions of big-game animals occurred over much of the globe. Thus, the great North American die-off was part of a larger phenomenon, one that related more to the end of the Ice Age, the Wisconsin glacial period, and subsequent climatic change than to human interference.

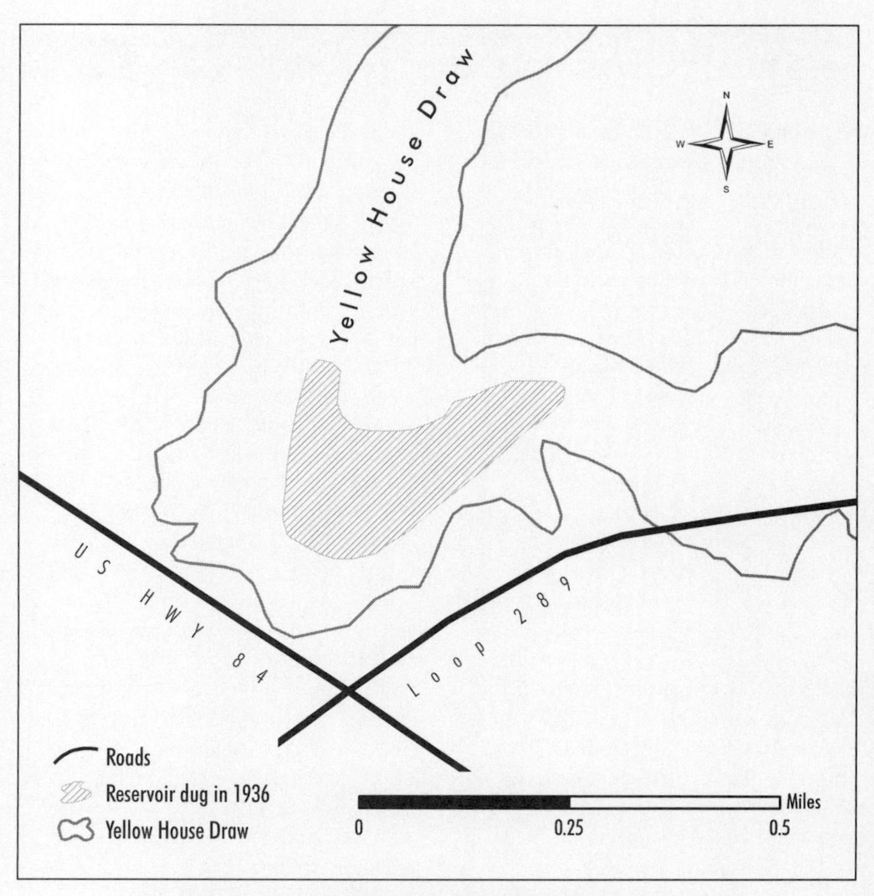

Yellow House Draw at Lubbock Lake Site

In such a view, climatic conditions shifted as the great glacial ice sheets retreated. In North America the nearly constant mild climate, regulated by the powerful influences of the huge glaciers, disappeared. While some places, such as Florida, remained wet, the arid Great Basin and the desert Southwest emerged.

On the Great Plains modern climatic patterns with seasonal extremes developed. Rainfall diminished, causing cold-water swamps, sloughs, bogs, and marshes to dry up. Thousands of playa lakes, which until then had contained permanent water supplies, shrank or dried up completely. A postglacial vegetation of short grasses replaced the tall, coarse grass savannahs with their scattered trees, thick forests, and numerous lakes that had supported mammoths and other large animals. With the large animals in decline, the continent-wide, nearly homogeneous Clovis culture faded into a variety of local traditions.

At the Lubbock Lake site and elsewhere, the Folsom tradition dominated after the decline of the Clovis culture. It extends from 10,800 years ago to about 10,200 years ago, and is characterized by its beautifully made, long-grooved—fluted—projectile points and by its heavy exploitation of now extinct large, straight-horned bison. Folsom-tradition people lived during the retreat of the glacial ice sheets, the changing climatic patterns, and the great ecological alterations.

The Folsom culture is named for a site near Folsom, New Mexico. In the early 1920s, George W. McJunkin, an African American who was foreman of a large cattle ranch in northern New Mexico, and a cowboy companion discovered the site. From the first McJunkin knew that the bones they saw were from some extinct animal. He informed interested people in Raton, New Mexico, and one of them, Charles Schwachevin, contacted Jesse D. Figgins of the Colorado Museum of Natural History. Figgins sponsored excavations at the site in 1926 and publicized his findings. By the end of 1927 Barnum Brown, paleontologist of the American Museum of Natural History; A. W. Kidder of the Peabody Museum; and Frank H. H. Roberts Jr., representing the Smithsonian Institution's Bureau of American Ethnology, had visited the site. Brown led excavations that produced bones from twenty-three or more large, straight-horned bison and nineteen projectile points. McJunkin, unfortunately, did not live to know that he had discovered the Folsom culture.

Folsom people had much in common with their Clovis predecessors. They were, of course, big-game hunters, but they concentrated on ancient bison rather than mammoths or mastodons, for the elephant-like animals had all but disappeared. They took elk, deer, and pronghorns. Like other big-game hunters, they also pursued many smaller animals, such as rabbits, raccoons, ducks, geese, squirrels, and several species of turtle.

Skilled stoneworkers, Folsom people produced the finest of fluted points. To make their distinctive projectile points they removed a single lengthwise flake from each side of the stone, thus producing grooves, or flutes, that ran from base to tip. The people used stone, bone, antler, and

Pleistocene Extinctions

During the last years of the late Pleistocene epoch, about fifteen thousand to ten thousand years ago, large numbers of animal species perished. In Europe the death of species was relatively small, but elsewhere around the globe late Pleistocene extinctions were massive. Perhaps 86 percent of the large animals of Australia became extinct, many of them long before fifteen thousand years ago. In South America large animal species suffered an 80 percent loss, and in North America the percentage of loss for large animal species may have reached beyond 66 percent.

Reasons for the great die-off at the Pleistocene-Holocene boundary remain something of a puzzle. A few theories describe such influences as disease epidemics, accidental change of sex ratios, astronomical agents, the beginning of agriculture, or other causes. Some of the most passionate, and most popular, arguments suggest overhunting by early humans, but the majority of the explanations cite changing climate and weather patterns at the end of the last great Ice Age—the Wurm-Weichsel-Wisconsin glaciation.

As the glaciers retreated, everyone agrees, the climate and weather patterns warmed, ecosystems changed, vegetation types shifted. Tall, coarse grass savannahs that supported large mammals, for example, disappeared in favor of short grasses and spreading forest lands. The tundra of northern Europe, Asia, and the Americas shrunk, also adversely affecting large mammals. Smaller mammals could adjust and move into empty ecological niches. The largest species, particularly sensitive to climatic change, had no such chance; they could move, certainly, but their grazing and foraging zones became smaller and spread over greater distances.

The scientists who maintain that humans played a significant role in the Pleistocene extinctions go further.

---

wood for other items, such as knives, choppers, drills, ornaments, and abraders, which were used for smoothing stone and wood objects.

As with the Clovis group, Folsom people, one might speculate, moved about in scattered bands of perhaps thirty to fifty people from several interrelated families. They followed ancient, now extinct bison across the Great Plains and doubtless came together in larger units in the summer for social and political reasons. They migrated in cyclical patterns, returning to favorite camping places and kill sites along shallow, marshy lakes, swampy meadows, and meandering streams. Although named for a site in New Mexico, good evidence of the unique culture is found at Lindenmeier, Colorado, a place to which Folsom people returned often to hunt bison. The Blackwater Draw sites contain Folsom materials as well.

At the Lubbock Lake site, vertebrate faunal remains from Folsom

Noting that late Pleistocene extinctions are not evident in the plant kingdom, they suggest that overhunting by humans represented a devastating impact on animal populations already in decline. Human overkill, as a result, became the difference between extinction and survival, and it may have contributed to the Clovis people's demise.

In North America, the argument goes, early hunter-gatherers came south from Alaska. They found a rich, mild environment that supported immense herds of bison and large numbers of other big game. Because of the abundant animal-food resources, the human population increased rapidly, and people made their way quickly through and across the steppe-like plains environment. As they moved, they hunted some animals, mainly large herbivores, into extinction. Ancient carnivores, dependent on the same big game, also perished.

Whatever the causes, in North America two-thirds of the large mammals present when humans first arrived disappeared. Paul Martin has noted that at least thirty-one genera of large herbivores vanished, thus placing pressure on carnivores, scavengers, and other creatures ecologically dependent on the grass-eaters. The ancient horse, two genera of ancient bison, and the ancient camel became extinct. Four-horn antelope, woolly and Columbian mammoths, mastodons, giant armadillos and ground sloths, giant anteaters, dire wolves, tapirs, saber-tooth tigers, short-faced bears, and many large rodents were among those animals that died out.

Source: P. S. Martin and H. E. Wright Jr., *Pleistocene Extinctions: The Search for a Cause* (New York: Yale University Press, 1967), v–viii, 77; Clive Ponting, *A Green History of the World: The Environment and the Collapse of Great Civilizations* (New York: St. Martin's Press, 1991), 33–35.

times suggest higher average annual temperatures with warmer summers than during the Clovis period. The site also exhibits extensive evidence of nearly constant Folsom occupation. Indeed, the discovery of a Folsom point there in 1936 led to the beginning of archaeological investigations at Lubbock Lake. Subsequent excavations revealed several Folsom points associated with extinct bison. Investigators have shown that Folsom hunters took bison—mainly females with calves—at Lubbock Lake in a series of smaller hunts, for no large trap existed there at the time. The hunters killed the animals around the margins of the site, which at the time contained a series of marshy-edged ponds. After the hunt, Folsom families processed their prey, gathered wild plants, and camped at the wet, marshy, lowland site.

In addition to bison at the Lubbock Lake site, archaeologists have dis-

This 1964 photo shows significant stratigraphic banding on the north bank at the Blackwater archaeological site. Mammoth bones dating to the Clovis period had recently been removed from the bank. Courtesy Blackwater Locality No. 1 Site archives, Eastern New Mexico University, Portales.

covered evidence dating to the Folsom period of several other mammals, plus amphibians, birds, and reptiles. They have found bullfrogs, green-winged teals, horned larks, and snakes. Among mammals, they have discovered hispid cotton rats, meadow voles, muskrats, coyotes, gray wolves, and others, along with evidence of a now extinct antelope. Folsom people probably hunted many such birds and mammals.

In hunting, Folsom people sometimes used the "surround," a hunting method that Clovis people had used with good effect, to drive their prey into a box canyon or a swampy spot of ground. Hunters on foot surrounded a small group of bison and herded them very carefully and very slowly toward the trap. Hunting bison on foot was dangerous, of course, but the animals could be moved about a mile or so without becoming overly agitated. In such a fashion the Paleoindians might kill a half dozen animals, thus providing meat for several weeks.

But Folsom hunters may have developed another hunting method, the bison jump-kill technique, in which they drove game over a small dropoff or cliff. Local variations of the method developed, but the technique, if in fact Folsom hunters first used it, continued to be used on the Great Plains until modern times. For a jump-kill operation to be successful, the hunters had to know the landscape intimately. Accordingly, they studied the land, learned its secrets, and used the knowledge to drive bison toward the dropoff. Favorite locations may have received repeated use.

Use of the bison jump-kill often ended in a massive slaughter. As the herd stampeded toward the dropoff—which might be only a foot or two high or a canyon edge—the lead animals, pushed from behind, could not stop. They tumbled over the precipice. If a high cliff was used as the dropoff, the bison probably died upon hitting the bottom of the cliff or from additional animals striking them from above. Some bison escaped. But many more of them than people could use suffered injuries, such as a broken leg or back. The hunters dispatched the crippled bison with their favorite weapons, spears and lances.

With a successful hunt over, butchering began. The hunters rolled the carcasses onto their backs, cut the skins down the belly, and pulled the hides down the flanks. In this way they preserved the hides for use in clothing, shelter, and tools. They then cut large portions of prime meat from the carcasses, taking the back, hump meat, fore and hind legs, shoulder blades, and rib cage. The butchers probably ate the tongues and some of the internal organs as they sweated with the heavy work.

Women likely participated in the butchering. Clearly they held important roles in Folsom society. They sometimes determined when the band should move camp, for example, and they took charge of the household. They probably also watched over the children, gathered plant foods, and cooked the meals.

As with the Clovis, the Folsom culture did not last. By 10,200 years ago it had begun to fade, and ultimately it disappeared. Ancient, straight-horned bison, the Folsom culture's characteristic food source, were in decline, and the mammoths, mastodons, ancient horses and camels, giant armadillos, short-faced bears, and many other Pleistocene animals had become extinct.

The environment was changing. Modern climate patterns were emerging. Temperatures were warming. Streams, lakes, swamps, and marshy lowlands were drying up. The plains grasslands were shifting to a mixed prairie cover of short grasses, such as grama and buffalo grasses. On the edge of the plains, forests—there probably were none on the Llano Estacado—were thinning and retreating eastward. Trees on the Llano were coming to exist only in the canyons and along the escarpments.

As the Folsom culture declined on the Great Plains, several regional traditions emerged. They represent the late Paleoindian period, dating between 10,000 and 8,500 years ago. The regional traditions are sometimes confusing and often obscure, but on the Llano Estacado they take such generalized names as Late Paleoindian, Plano, Firstview, and Plainview. Plainview-age remains often exist at sites that Folsom people had occupied. On the Llano Estacado, there are as many Plainview sites as Folsom and Clovis sites combined.

Projectile points usually distinguish Plainview groups from Folsom and Clovis traditions. Plainview points were often long, slender, and nearly parallel-sided with a thick center and a flat base. Plainview points and knife blades range from two and a half to five inches long, but they vary from site to site. Thus, the terms *Plainview, Firstview, Plano,* and *Late Paleoindian* serve to identify a whole series of traits and artifacts with no agreed-on class name.

Plainview-style points first appeared in quantity and context at a site in Plainview, Texas. The site, in a wide, meandering stretch of Running Water Draw in the upper Blanco River, became known in 1944 when contractors moved in to dig a gravel pit. Although local inhabitants had collected projectile points there for several years, the first professionals to investigate the site included Elias H. Sellards, director of the Bureau of Economic Geology and the Texas Memorial Museum of the University of Texas; his colleague Glen L. Evans; and Grayson Meade of Texas Technological College.

The evidence at Plainview, the three archaeologists reasoned, showed that hunters drove large, ancient, straight-horned bison over a short dropoff or small cliff and then killed the crippled animals that could not escape. People used the kill site on more than one occasion. Bones from about one hundred bison and twenty-eight projectile points have been found at the site. Other artifacts exist in close association with the bone beds.

But archaeologists found neither skulls nor tails among the bone beds. Clearly, the Plainview hunters carried off the heads. And, apparently, they kept the tails attached to the hides that they had removed from the carcasses, using a butchering process similar to what Folsom people had used. Perhaps they hauled the heads to a nearby hide-tanning site, where they bashed open the skulls to get the brains for use in tanning hides, something modern Plains Indian people did. Or they may have used the skulls for other, or additional, uses, such as spiritual altars.

Plainview people were big-game hunters, of course, but they did not hunt mammoths or mastodons. The elephant-like mammals had disappeared by that time. Bison were the culture's characteristic prey, but many Plainview-age kill sites suggest that the people also took deer, elk, and pronghorns, and they sought many smaller animals, including rabbits, raccoons, turtles, and birds.

On the Llano Estacado, Plainview sites can be found at several places, including Plainview, Blackwater Draw, and Lubbock Lake. Sites at Milnesand, New Mexico, Lake Theo, and elsewhere across the high tableland have revealed slight differences in projectile construction and design. The changes suggest that Plainview-age groups did not possess the relative cultural homogeneity that characterized Clovis people or, to a lesser extent, Folsom groups.

Nonetheless, the lifeways of the Plainview people were much like those of their Clovis and Folsom ancestors. Plainview people employed the jump-kill method for bison hunting, at least at places on the Northern Plains, and the entrapment method. Their hunting schemes suggest that Plainview groups possessed more complex sociopolitical organizations than other Paleoindian people, which means larger bands and more frequent interband meetings. They depended more on seed collecting and root digging than their forbearers, suggesting the development of a varied diet on the Great Plains.

Plainview people may also have worked out an important new way to preserve meat. It was a method that foreshadowed modern Plains Indians' pemmican. In the new process, women dried small strips of meat in the sun, pounded the dried meat into small bits, mixed it with animal fat and berries, and packed the concoction in skin bags that they then sealed with fat.

At Lubbock Lake, Plainview artifacts date to about ten thousand years ago. Plainview peoples, including groups called "Plano" and "Firstview,"

The Plainview Bison Kill Site

Located along Running Water Draw near downtown Plainview, Texas, the excavation site provided a name for the Plainview culture. The site featured a large bed of bison bones with projectile points among them. People in Plainview discovered the archaeological value of the site after a contractor began mining the place for Ogallala caprock, or caliche, used for gravel.

The Plainview site, located where Running Water Draw was once a wide, marshy stream that meandered gently through the area, was locally known as a favorite spot for amateur point collectors. First investigated systematically in 1944 by Glen L. Evans and Grayson E. Meade, the site was a place where Indian hunters either ran bison over a small, steep cliff or trapped them in a muddy bend of the stream. Archaeologists found bones to more than one hundred bison, at least twenty-eight projectile points, and other significant lithic artifacts throughout the site.

Recent interpretations of the evidence suggest that about seven thousand to nine thousand years ago hunters used the site repeatedly, apparently for fall and spring hunts. Archaeologists found no skulls or tailbones, suggesting that the people who hunted the site removed the bison heads and hides—tailbones intact—for processing elsewhere. The hunters may have saved the skulls for religious purposes; the hunters also may have cracked open the skulls to use the brains during hide tanning.

Researchers dug pits at the site to study the site's stratigraphy and to examine the bone beds. They enlarged the first pit in 1945, and in subsequent years dug two additional pits. In the 1970s the first pit became a garbage dump, destroying some of its research value. Later, Eddie Guffee, associated with the Wayland Baptist University Archaeological Research Laboratory, found some of the bone bed intact east of the pit. The other pits provided important geological and stratigraphical information before they too became buried under garbage.

The Plainview site has not been investigated systematically in nearly thirty years.

Eddie Guffee's excavation in 1977 at the Plainview bison jump site in Running Water Draw. The systematic dig began after removing nine feet of fill. Courtesy Eddie Guffee.

used the wet, marshy, lowland site to camp and to waylay bison and other prey coming to drink. They did not stay permanently at the place but returned often to hunt. Perhaps they came ahead of the bison herds, knowing the animals would not be far behind. And, like the Clovis and Folsom peoples before them, they watched, studied, and followed their chief animal food source.

At the Lubbock Lake site, archaeologists have discovered several vertebrate faunal remains dating to the Plainview and Firstview periods. The remains include fish, such as bullheads; amphibians, such as tiger salamanders, toads, and leopard frogs; reptiles, such as yellow mud turtles, box turtles, and Texas horned lizards; and birds, such as pied-billed grebes, mallard and pintail ducks, and teals. They have also found remains of cottontail rabbits, blacktail jackrabbits, blacktail prairie dogs, plains pocket gophers, wood rats, muskrats, coyotes, gray wolves, badgers, pronghorns, and now extinct bison.

Also at the Lubbock Lake site, Alibates flint and Edwards Formation chert dominate Plainview points, knives, and tools. Small bands of people traveled to the Texas Panhandle to work the Alibates quarry there. Others probably traded for the popular flint. Edwards Formation chert, a fine-grained material that varies in color from tan and gray to dark blue, exists at many outcrops along the Llano Estacado's eastern escarpment, but it can be found across a vast swath of central Texas. Tecovas Jasper—the mottled red, tan, yellow, and white Quitaque flint—was also present along the eastern escarpment. Thus, with a short trip to the Llano's eastern edge, or through trade, people might acquire good, workable stone. As with the Clovis and Folsom groups before them, the Plainview people reused broken or damaged projectile points as knives or scrapers or other tools.

In the upper Brazos River region and at the Lubbock Lake site, the Paleoindian period lasted about three thousand years. Its people were hunters and foragers who moved across large territories in search of the animal and plant foods that typified their diets. In a general sense during the long span of time, three broad cultures emerged: Clovis, Folsom, and Plainview, with the latter best identified as a whole series of closely related traditions.

Weapon points and various stone tools distinguish the Paleoindian cultures from one another. Lost or discarded points, knives, scrapers, and other tools and weapons have been found at sites over much of the Llano Estacado. The stone points and tools suggest that humans occupied some of the sites, such as Lubbock Lake and Blackwater Draw, at various times throughout the Paleoindian period.

Food supplies and hunting techniques evolved through the period. In the years of the Clovis people, Columbian mammoths; ancient bison, horses, and camels; and other extinct species were part of the diet, but the people also took rabbits, rodents, turtles, and birds. Hunting strategies were unsophisticated with small groups seeking to ambush single animals. In the middle years, which were dominated by Folsom hunters, ancient bison formed a larger share of the diet, but Folsom people ate deer, elk, pronghorn, and smaller game as well. They adopted large-group hunting strategies, such as the bison jump-kill technique, and, as reflected in their hunting schemes, organized into more complex sociopolitical forms than the Clovis groups. Plainview people of the late Paleoindian period saw bison dominate their food supply, but the people also consumed smaller animals, including ducks, geese, turkeys, rabbits, and muskrats.

Various archaeological sites, especially those at Lubbock Lake and Blackwater Draw, suggest a number of conclusions. Paleoindian hunters maintained a low population density. They traveled in a yearly pattern from campsite to campsite in a range of territory that covered perhaps hundreds of square miles. They moved about in small bands of interrelated families. Families were small and, as they were hunting societies, probably male dominated with descent determined by the male line. Men hunted. Women and children gathered edible plants, nuts, and roots. When food sources were plentiful, life for the Paleoindian hunting bands was relatively easy, allowing people to rest and spend their time as they pleased.

By 8,500 years ago the Paleoindian period had begun to give way to new traditions. The change was partly in response to shifting climatic patterns that had altered the environment and produced a hotter, drier Great Plains. People began to cultivate some of the plants they had been gathering, and they stayed longer in one place. Over time such events in the upper Brazos River region and at the Lubbock Lake site, even as they ended Paleoindian traditions, ushered in the long, dry Archaic period.

★

Chapter 3

The Archaic Period

Living Through Drought

ON THE LLANO ESTACADO and at the Lubbock Lake site the Archaic period extends from about 8,500 years ago to 2,000 years ago. Characterized by long, dry years of extensive drought interspersed with times of higher rainfall, the 6,000-year period underwent changes in its climate and weather, in its plant and animal life, and in its patterns of human occupation. For the Llano Estacado, archaeologists subdivide the Archaic into three chronological parts: early (8,500 to 6,500 years ago), middle (6,500 to 4,500 years ago), and late (4,500 to 2,000 years ago).

In many ways the shift from Paleoindian to Archaic was a gradual one. But, still, an abrupt modification in projectile-point forms marks the changeover. Clovis and Folsom peoples produced lance-shaped points with center grooves, or flutes. Early Archaic people left out the center groove, but they cut side notches at the base of their points. They used the notches and tiny rawhide strips to secure the point to the spear or arrow shaft. By the middle Archaic, additional point innovations had occurred: points became smaller, the notches more pronounced. Such innovations continued toward the late Archaic, when a variety of corner-notched points with wide, open notches that formed sharp points appeared.

Unlike Clovis societies, which possessed some continent-wide manifestations, Archaic societies included many local adaptations. During the Archaic period, as the massive northern ice sheets continued to retreat, distinctive new environments, such as a humid Southeast and an arid Southwest, emerged. In such regions Archaic peoples adapted to the changed conditions and, in effect, adopted subsistence patterns related to the ecological zone, or zones, in which they lived. In other words, regional specializations came into existence. Archaic people in the eastern woodlands, for example, developed cultures and societies different from the people along the Northwest Pacific Coast.

In a related sense, some scholars divide the early Archaic into three broad geographical regions, or cultural areas. The eastern area included territory from the Atlantic Ocean westward to just beyond the Mississippi River, but it also extended up the Red and Arkansas river basins and into East Texas. In this forested region of plentiful rainfall a woodland culture developed. The western area reached from the Rocky Mountains to the Pacific Ocean. In this dry, arid upland of sparse vegetation something of a desert culture developed. The third principal area was the Great Plains, spreading west from the Mississippi to the Rockies. In this steppe-like environment of few trees but extensive grassland that included the upper Brazos River country and the Lubbock Lake site, a bison-hunting, foraging culture had developed.

In all three geographical regions, Archaic people were hunters and foragers. Like the Clovis and Folsom groups before them, their subsistence techniques, tools, and utensils focused on food collecting rather than on food production through horticulture. Game, of course, was always important, but vegetable foods became the basis of the Archaic diet. The people preferred fruits, berries, nuts, tubers, and seeds, all of which they sought on a scheduled round of collecting.

Archaic groups in the Northeast foraged for plant foods that were local variations of the nuts, tubers, and berries that people in the Great Basin ate. The same is true for game. Deer, for example, became a major part of diets in the Northeastern woodlands. People in the Great Basin sought rabbits, sometimes through large community hunts. People on the Great Plains, including the upper Brazos River region and its Lubbock Lake site, hunted bison as their major meat source.

In the foreground of this 1964 photo several people are excavating on the south bank of the Blackwater Draw site north of Portales, New Mexico. In the background, a sand and gravel company's buildings and equipment can be seen at the Blackwater Locality No. 1 site. Gravel operations are no longer conducted at the site. Photo courtesy Blackwater Locality No. 1 site archives, Eastern New Mexico University, Portales.

Thus, selective and seasonal exploitation of resources characterize the Archaic. Variety, inventiveness, and adaptability also characterize it. The period represented a whole series of richer, more versatile, and more technologically advanced cultures than the Paleoindian, big-game-hunting groups.

For the Archaic, archaeologists find a wider range of tools and implements used for a greater variety of purposes than what was employed by earlier peoples. Archaic sites reveal stone points and knives used for hunting and butchering. In the eastern woodlands sites, however, and along some of the western rivers, one might also find nets and hooks for fishing.

The most common tools and utensils of the Archaic era reflect a reliance on plant foods. Baskets and skin containers for collecting seeds and nuts were plentiful. Mortars, pestles, and sandstone grinders occur. Roasting pits, some of which the people used repeatedly, possibly for centuries, became common. Archaeologists excavating the Lubbock Lake site have found such an oven dating to the middle Archaic.

American Bison

Ancient bison colonized North America before the Wisconsin glaciation, which extended from seventy thousand to ten thousand years ago. The older species, some of which may have arrived as early as six hundred thousand years ago, were huge, with great horns that reached six or more feet from tip to tip.

The archaeological and paleontological records suggest that in North America three principal bison species occurred: *Bison latifrons* (now extinct), *Bison antiquus* (now extinct), and *Bison bison* (the modern American buffalo). Among the latter two, subspecies evolved.

In North America perhaps the earliest species was *Bison latifrons,* an enormous animal with great horns. Twice the size of modern bison, *latifrons* spread from Florida to California and as far south as central Mexico, perhaps farther. The long horns provided a means to fight off carnivore predators such as dire wolves and saber-tooth tigers or a means of aggressive display toward large but nonhorned mammals on the Great Plains, including mammoths, tapirs, and giant ground sloths.

Bison antiquus descended from the long-horned *latifrons.* They evolved smaller—although still relatively large—and slightly curved horns that are on the order of horns on modern bison. Having decreased in body size over a period of about fifty thousand years, they were smaller than their ancestors, but still remained much larger than modern bison. They possessed a heavy, distinctive hairy mane, a heritage from their evolution in the far north. Apparently it was adequate as a means of display for aggressive purposes. Like their ancestors, *Bison antiquus* thrived in North America and spread from coast to coast and as far south as central Mexico. On the Llano Estacado, Clovis and Folsom people hunted a subspecies of *antiquus.*

Modern bison—*Bison bison*—most scientists argue, evolved from one of the subspecies of *antiquus: B. antiquus occidentalis.* Older forms, such as *Bison antiquus,* were Ice Age animals that possessed a short, spring calving period so that the young could mature enough to survive their first cold, harsh winter. Such post–Ice Age bison as *B. antiquus occidentalis* and modern bison evolved a longer breeding season and gave birth to small calves. They adapted well to the short-grass plains, became less selective in eating (they were both grazers and browsers), and saw their winter mortality rates decline. Modern bison flourished on the Great Plains, even as other Pleistocene mammals, including *B. antiquus occidentalis,* came under increasing stress.

Bison belong to the family *Bovidae.* The American bison, or buffalo as we call them, are classified as genus *Bison* and species *bison bison,* making their full name *Bison bison bison.* Among large mammals Bison represent a recent genera, having first appeared in Asia during the late Pliocene (5 million to 1.8 million years ago). They are part of the oxlike bovids—the Bovini. *Bison* lived through much of Europe, Asia, and North America, the only Bovini genus to gain such widespread distribution without the assistance of humans.

North American Native Species

- *Bison latifrons*
- *Bison antiquus*
  - *Bison antiquus antiquus*
  - *Bison antiquus occidentalis*
- *Bison bison*
  - *Bison bison bison*
  - *Bison bison athabascae*

Source: Jerry N. McDonald, *North American Bison: Their Classification and Evolution* (Berkeley: University of California Press, 1981), 57.

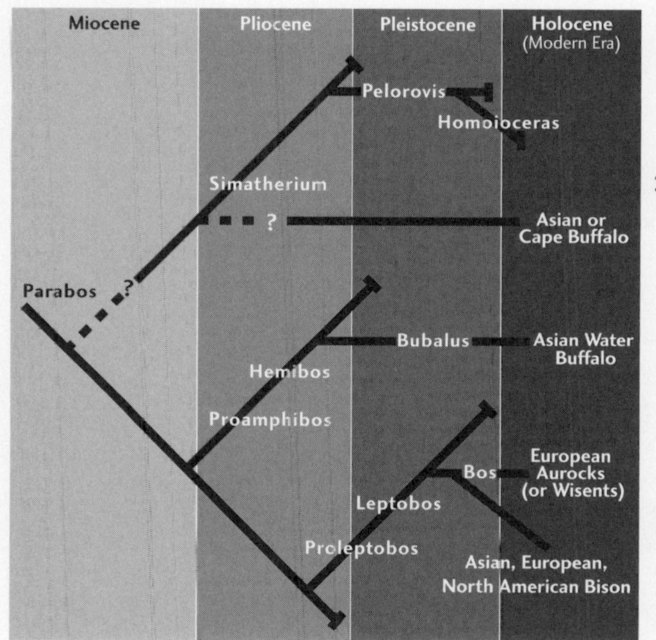

Bovini Phylogenesis

Ancient bison as depicted in the outdoor, life-size sculpture at the Lubbock Lake Landmark. Photo by Tony Gleaton. Courtesy Southwest Collection.

Most sites reveal tool kits that include axes, adzes, scrapers, and gravers. Such tools might have been used to manufacture weapons and implements of wood or bone. The sites also reveal ornaments of bone, teeth, and shell. In some places Archaic people turned turtle shells into fashionable ornaments.

On the Llano Estacado only a few Archaic-era sites have been excavated. Archaeologists have noted a large number of them, but geologic conditions during the Archaic did not serve site preservation well. Nonetheless, at the Lubbock Lake site and on the Llano Estacado, the archaeological record is revealing. The record suggests the development of the extreme upper Brazos River system. Through thousands of years the major tributaries of the upper Brazos River had been cutting into the Llano Estacado. Water draining eastward off the level plain was respon-

The Altithermal

Altithermal is a geological and climatic designation for an extensive period of drought that occurred across the Great Plains during the Archaic period. Varying degrees of severity, both geographically and climatically, characterized the Altithermal. In the upper Brazos River region and at the Lubbock Lake site it extended from about 6,500 years ago to about 4,500 years ago, roughly the years of the middle Archaic period. During the Altithermal, dry conditions and high temperatures on the Great Plains reached maximum limits. A few scientists suggest that the event caused humans to abandon some of the drier places on the plains.

Obviously, then, the Altithermal brought changes. Tallgrass prairie gave way to shortgrass steppe, and in the most severe instances the short grasses tended to give way to desert shrubs and near-desert conditions. Some places became nearly bare of vegetation, and, accordingly, a few scientists argue that during the long Altithermal a Great Basin desert culture emerged on the Great Plains. Wind picked up soil from the thinning ground and deposited the material in dunes, draws, and valleys. Water tables dropped and many springs quit flowing. Grasses thinned. Bison and other large mammals on the plains declined in number as they sought drought refugia elsewhere.

In response, Archaic peoples on the plains turned more and more to plant foods. Where water remained, such as in larger streams, they may have turned to aquatic resources. They probably tended to stay near streams and rivers that held permanent water sources. Lubbock Lake site, for example, held water during the Altithermal, and, the archaeological record suggests, supported a human population throughout the long drought.

Causes for the Altithermal are not precisely known. Increased sun spot

sible. Over time the runoff created deep canyons along the Llano's eastern edge. About ten thousand years ago, at the Pleistocene-Holocene, or Modern era, boundary, the far upper draws began to form with runoff water helping to create deeper cuts at the Lubbock Lake, Plainview, and Clovis sites. The extreme upper Yellow House, Blackwater, and Running Water draws usually contained water. But over the past couple of centuries the extreme upper ends of Yellow House and Blackwater draws have begun to fill with dirt, sand, and dust.

The archaeological record further reveals that in the early Archaic period, 8,500 to 6,500 years ago, some ancient bison (*Bison antiquus*) still existed on the Llano Estacado. Clearly, the species was in decline, and, as suggested by size diminution, it was also experiencing severe stress. At the Lubbock Lake site an ancient bison bone bed shows that humans of the

activity is one theory. Most theories suggest changes in atmospheric air masses and wind patterns. Three air masses characterize much of the interior of North America: the Arctic, the Tropical Maritime, and the Mild Pacific. The dry Mild Pacific air mass dominates the plains for about 50 percent of the year and probably accounts for the grassland environment. It occurs through much of the winter, resulting in low winter precipitation across the plains, and part of the summer, times when the shortgrass plains receive very little rainfall. Arctic and Tropical Maritime air masses carrying greater precipitation intrude on the Mild Pacific in the late winter and early summer, bringing rain in the spring (May–June). The Tropical Maritime carries rain to the tallgrass prairies through much of the summer and into the fall.

Seasonal changes in the three air masses, governed in part by the waxing and waning of the upper tropic and subtropic jet streams, impact rainfall amounts. When the Mild Pacific intrudes into the Arctic or the Tropical Maritime, no rainfall might reach the Great Plains. A year-round dominance of the Mild Pacific results in severe drought, eastward extension of shortgrass plains and climate, and the subsequent displacement of the tallgrass prairie and climate. In sum, if Arctic and Tropical Maritime air masses cannot bring spring rains to the shortgrass plains, trouble follows: grass thins, forage yields decline, and the carrying capacity of the range land decreases. Large herbivores must leave or die. Again, causes for the long Altithermal period remain speculative, but a centuries-long shift in the Arctic, Tropical Maritime, and Mild Pacific air masses may have played a key role.

Source: Brian O. K. Reeves, "The Concept of an Altithermal Cultural Hiatus in Northern Plains Prehistory," *American Anthropologist* 75 (1973): 1221–53; Vance T. Holliday, "Middle Holocene Drought on the Southern High Plains," *Quaternary Research* 31 (1989): 74–82.

early Archaic period killed three such animals there. Unfortunately, scientists who excavated the site have found few material artifacts.

But changes came. Modern plains bison (*Bison bison*), with curved and shorter horns than the older species, were evolving. They evolved from a mixing of the large, ancient, straight-horned *Bison antiquus* and a smaller, postglacial species, *Bison occidentalis,* that had invaded the plains. Modern plains bison adapted to grazing on both shorter and less coarse grasses that were spreading through the plains and adjusted to the warmer, drier climate. Females among them, although developing a longer gestation period, began giving birth at a younger age than the larger varieties. In short, bison adjusted to the changing environment by growing smaller and giving birth earlier, a not uncommon reaction among animals under survival stress. North American cod and fish of the Mediterranean Sea have undergone similar changes.

The archaeological record also shows how the climate changed. The Archaic witnessed a warming trend that saw its environment shift from the cool, wet features of the Paleoindian period through a long stretch of warm, dry years—about two thousand years of drought—before a return to greater moisture.

The long dry spell occurred in the middle Archaic period, from 6,500 to 4,500 years ago. Scientists who have investigated the Lubbock Lake site refer to the long drought as the local manifestation of the *Altithermal,* a semiarid period of little moisture on the Great Plains and of warmer and drier conditions than had come earlier. During the Altithermal, bison and other animals, dependent on reliable water sources, declined in numbers on the Llano Estacado. They left to seek drought relief in regions to the east. Human groups, however, lived on the Llano through the Altithermal, and water, although in declining amounts, was present in the major streams throughout the period.

Plants surrounding the Lubbock Lake site changed during the Archaic period. Species appeared and disappeared before returning again in response to the Altithermal and its drier, warmer conditions. With subsequent decreases in moisture and humidity came a gradual decrease in the vegetative cover. Eventually, during the Altithermal of the middle Archaic, the treeless dry prairie surrounding the Lubbock Lake site became something of a desert plains grassland. Range conditions deteriorated as heat-

and drought-tolerant desert scrubs replaced the thick plains grasses. Dusty, sandy conditions became common.

In Yellow House and Blackwater draws during the Altithermal the marshy conditions also changed. With dry conditions common and grass cover thinning, the Llano Estacado's constant winds picked up soils and blew them onto the valley floors. Water remained at the Lubbock Lake site, but the freshwater marsh became more alkaline. Marsh plants continued to provide food for both humans and animals.

Indeed, for grassland animals, Blackwater Draw near modern Portales, New Mexico, and Yellow House Draw at modern Lubbock served throughout the Altithermal as refuge from drought. But with range conditions rather poor for large plains animals, bison herds decreased in size. Perhaps the thinning herds account, in part at least, for humans turning more and more to a reliance on plants for sustenance.

Still, people used the Lubbock Lake site throughout the Archaic period, including the Altithermal. They hunted bison and other game in the draw. They camped and gathered plant foods there—even during the harshest years of the Altithermal. They gathered such water-based foods as shellfish. During wet years, as at the end of the Archaic, when the amount of water at the site increased, they may have used nets or hooks to catch catfish.

In any case, survival through the long drought of the Altithermal depended on diverse subsistence schemes. And, indeed, Archaic people adopted a variety of food-gathering strategies. The people shot waterfowl, hunted deer, and foraged for many species of wild plant foods, including prairie turnips, groundnuts, cactus fruits, sunflowers, Jerusalem artichokes, and seasonal fruits. They paid close attention to the changing grazing conditions on the plains grasslands and carefully monitored bison movements, relying in the hunt on their own patience and their highly developed tracking skills. Communal kills of bison and other game became common.

In the late Archaic period, 4,500 to 2,000 years ago, the climate turned wetter and probably a bit cooler. Yellow House Draw also changed. Near the end of the period, it was again a shallow, wet, marshy valley. The draw was also wide. At the Lubbock Lake site, where its prominent and characteristic horseshoe-shaped bend occurred, the draw contained a wet

Selected Vegetation Types

Forest	Trees dominate and are closely spaced; usually exists with humid climate and the canopy is complete
Woodland	Trees or large shrubs dominate; canopy not complete
Parkland	Large openings in forests or woodlands; trees and shrubs present, but herbs dominate
Brushland	Smaller shrubs dominate; canopy not complete
Savanna	Trees, shrubs, and herbs present over an extensive area; on Great Plains usually associated with semiarid or dry climates; tall grasses are present
Grassland	Grasses dominate; trees and shrubs are rare; on steppe-like Great Plains short grasses adapted to arid or semiarid environments prevail; in prairies tall grasses prevail; ground cover is usually complete.
Tundra	Herbs, mosses, and lichens dominate; located where cold weather exists with short warm seasons; shrubs and small trees might be present
Desert	Exists under arid conditions; ground cover not complete; shrubs and herbs dominate

Source: Adapted from Jerry McDonald, North American Bison: *Their Classification and Evolution,* Berkeley: University of California Press),16.

meadow of mixed grasses. Good, fresh water existed at the small reservoir or lake, and an active spring above the bend and along the west valley wall about six hundred feet from the main stream provided additional water. Some trees, particularly hardwoods such as netleaf hackberry, appeared or perhaps reappeared after the long Altithermal.

In many ways the climate resembled our present-day weather. In a general sense, then, a modern climate interspersed with short periods of greater aridity or drought has existed in the upper Brazos River region and at the Lubbock Lake site for the past 4,500 years.

Geologically, the Archaic period witnessed the creation of various soils. Soils dating to the early Archaic reflect the shifting climate patterns. At the Lubbock Lake site, for example, the geology for the period displays windborne sediment deposits in what archaeologists call the Firstview Soil. After the initial deposits, Firstview Soil developed from sediment deposits related to a wet, marshy surface. Thus, between 8,500 and 6,500 years ago a dry climate that favored windborne depositions of sand, dirt, and dust preceded a time of relative moisture that favored wet, marshland deposits of mud and claylike materials. Firstview Soil, which took about two thousand years to form, is the oldest soil at the Lubbock Lake site.

In the late Archaic, Lubbock Lake Soil, a well-developed and widespread soil of sandy materials mixed with claylike marshy deposits, first appeared. The soil suggests that during the late Archaic a lowland marsh existed at the Lubbock Lake site. Over time the marsh alternately advanced and retreated. The land surface remained stable, for the most part, and water could be found in the marsh.

Scientists working through Archaic-period soils and deposits at the Lubbock Lake site have found evidence of many animals. Among amphibians, they have found salamanders, toads, bullfrogs, and leopard frogs. They have found evidence of such reptiles as mud and box turtles, lizards, and snakes of several kinds: garter, king, patch-nosed, hog-nosed, bull, and whip snakes. Clearly, birds were present, but archaeologists have not recorded many species. They have noted the presence of jackrabbits, prairie dogs, gophers, mice, several species of rats, coyotes, pronghorns, and bison, both ancient and modern species.

Selected Fauna of Lubbock Lake Site

Periods in Which Species Have Been Found

	Clovis	Folsom	Plainview/ Firstview	Archaic	Modern
MAMMALS					
desert cottontail	X		X		
cottontail		X	X		
black-tailed jackrabbit	X	X	X	X	X
black-tailed prairie dog	X	X	X	X	X
plains pocket gopher	X	X	X	X	X
pocket mouse	X	X			
white-footed mouse		X	X		
hispid cotton rat	X	X	X		
meadow vole	X	X	X		
prairie vole	X	X	X		
muskrat	X	X			
coyote	X	X	X	X	X
gray wolf	X	X	X		X
kit fox	X				
badger			X		X
bobcat			X		

Selected Fauna of Lubbock Lake Site *(continued)*

	Clovis	Folsom	Plainview/ Firstview	Archaic	Modern
Columbian mammoth	X				
stilt-legged horse	X				
horse	X				
peccary	X				
camel	X				
extinct antelope		X			
pronghorn		X	X		X
extinct bison	X	X	X	X	
modern bison				X	X
BIRDS					
ducks	X	X	X		X
turkey	X				
Canada goose	X				
snow goose	X				
grebe			X		
grouse			X		
gray-breasted crake			X		
coot			X		X
mountain plover			X		
burrowing owl	X				
common nighthawk		X			
northern flicker		X			
red-winged blackbird	X		X		
cowbird			X		
vesper sparrow	X				
AMPHIBIANS					
bullfrog	X	X	X	X	
leopard frog	X	X	X	X	X
toad	X		X		
cricket toad	X				X
plains toad	X				X
tiger salamander	X		X	X	
REPTILES					
snapping turtle	X	X			

	Clovis	Folsom	Plainview/ Firstview	Archaic	Modern
yellow mud turtle	X	X	X	X	X
carolina box turtle	X				
ornate box turtle	X	X	X		X
elegant slider	X	X	X		X
soft-shelled turtle	X				
hog-nosed snake	X			X	
king snake			X	X	
milk snake	X				
coach snake			X		
water snake	X				
copperhead			X		
rattlesnake	X				X
Texas horned lizard	X		X		X
FISH					
catfish	X				
garfish	X				
quillback	X				
bullhead	X	X	X		
white bass		X			
sunfish		X			
warmouth		X			
logperch		X			

Source: Eileen Johnson, "Zooarchaeology and the Lubbock Lake Site," in *History and Prehistory of Lubbock Lake Site,* ed. Craig C. Black, 114–15 (Lubbock: *The Museum Journal* 15, 1974); Eileen Johnson, "Vertebrate Remains," in *Lubbock Lake: Late Quaternary Studies on the Southern High Plains,* ed. Eileen Johnson, 49–65 (College Station: Texas A&M University Press, 1987).

Not much is known about Archaic-period social organization, at least for the Llano Estacado. Nor is there good information about kin and marriage ties that linked neighboring bands. However, sharing of work and food sources, egalitarian political and interpersonal relationships, and social flexibility must have been present. That does not mean that everyone was equal, of course, for some men were better hunters than others and some were better stone knappers or more effective leaders.

One might speculate that on the Texas High Plains, Archaic people moved around in small bands in a highly mobile lifeway. In their annual treks they covered a wide area. They came together in larger numbers, perhaps four or five bands, in the summer when bison, during the mating season, also congregated in sizable numbers. The larger aggregations of humans provided opportunities for finding marriage partners, renewing extended family ties, and exchanging ideas and material items, such as stone blades, points, and knives.

Because the groups were hunting societies, men probably governed family organizations. Men hunted, made tools and implements, fashioned weapons, and chipped away at stone projectile points. Because Archaic people still relied on the hunt for sustenance and because hunting was a male occupation, men probably dominated decision making, but such a conclusion is speculative. The bands were patrilineal and patrilocal. Thus, men remained with the kin group into which they were born, and women married into neighboring groups.

That is not to say that women were inconsequential in each band's social and political organization—far from it. With men often absent from the band on a hunt, women needed to control camp life and govern daily decision making. They often determined when it was time to move camp and seek out old trading partners. They took charge of the household, prepared the meals, watched over the smaller children, and fashioned the clothing. They gathered food to eat and collected wood or bison dung for fires.

Women most likely made the family's portable house as well. Using bone needles and sewing awls, they produced skin tents of bison hides—lodges or "tipis" we would call them at a later date. About six feet in diameter, the tents were small, mainly because family members needed to carry them when moving camp. Near the end of the Archaic, some women pressed dogs into service; they fashioned pole frames to the dogs and placed the tents on the frames. The dogs then dragged the tents. Women may have placed other small loads on the backs of dogs for the animals to carry.

Women may also have made pemmican, a food that proved an effective way of preserving meat. Pemmican is produced by pounding sun-dried strips of meat into fine pieces and mixing the meat with melted fat, bone marrow, suet, and sometimes dry paste from wild fruit, especially berries, cherries, or plums. For flavor the women may have added nuts. They stored the resulting product in skin bags, sometimes sealing the bags with melted tallow. Pemmican may keep for years.

Population growth on the plains was slow. It occurred, but it probably expanded and retracted in relation to the number of bison present. Thus, human population probably thinned during the Altithermal but expanded during the late Archaic when bison herds increased dramatically.

Clearly, changes in bison and other large game—their numbers, their disappearance, their reappearance—directly affected people's lives. In part the impact may have been spiritual; Archaic people, like early hunters elsewhere in time and place, probably sought spiritual connections with the animals they hunted. They designed any manner of ritualistic efforts to gain the favor or cooperation and forgiveness of their prey. As a result, perhaps, Archaic people, upon noticing the thinning herds, believed they had transgressed and somehow offended bison—or, perhaps not.

Still, belief systems and spiritual practices centered around the Archaic peoples' close relationship to nature and the animals they hunted, especially bison. They mimicked the bison, wore bison headdresses, and before major hunts held dances and sang songs to bring their prey closer. They may have utilized bison-skull altars, prayed to the animals, thanking their dead prey for giving themselves up to the people. In that sense, hunting was as much a spiritual activity as an economic one.

Among most hunting bands, including probably the Archaic groups, no priestly class existed. Shamans, who could be men or women, might direct the spiritual activity that was group oriented, such as prehunt dances or posthunt thanksgivings, but religion was largely an individual affair. Archaic people, as with most people living close to nature, probably saw their sacred world as populated by many forces and spirits, and, believing in physical and mystical connections among them all, they honored the spirits, the land, and the animals. Like many hunting bands, they probably accepted the idea that all living creatures had souls and talked to people. However, for the peoples living on the Texas High Plains during the Archaic period, researchers can only speculate about these beliefs. There are few pictographs or similar evidence available for the period.

On a more practical level, Archaic people, when animal numbers declined, sought alternative food sources. As indicated above, they devoted more time to collecting food, gathering seasonal fruits and nuts, and developing a more diverse commissary. Not clear is their concern for the environment or the related idea of protecting food sources for the next generation. In part the answer is a positive one, for some Archaic people, although probably not those who camped at the Lubbock Lake site, began planting the seeds of some of the wild foods they gathered.

As time passed Archaic life patterns gave way to more sophisticated practices. Social organizations became more complex, for example, and nonmaterial cultural traits grew in importance. People took to using tobacco and spending more time on handicrafts. Off the plains toward the east, where large burial mounds appeared, people developed a greater concern for those who had died. In the desert Southwest horticultural village life replaced Archaic hunter-gatherer lifeways. Although many features of the Archaic tradition continued in use for several more centuries, particularly on the Great Plains, life over much of the American continent underwent a gradual adaptation to other subsistence modes.

In the American Southwest a communal tradition that centered on horticulture emerged. Characterized by irrigated horticulture, hunting of small game, gathering of nuts and seeds, the use of pottery, and a close adaptation to the desert environment, this tradition began about 2,800 years ago, near the end of the Archaic period.

About the same time, in the region east and just west of the Mississippi River and in East Texas, a Woodland tradition emerged. Limited horticulture, hunting of deer and small game, gathering of fruits and nuts, the use of pottery, and mound building characterized it.

More than that, however, both in the East and in the Southwest distinct cultural communities, having evolved from Archaic traditions, eventually appeared. In the East they included the Adena, Hopewell, and Mississippian societies. The Adena, which concentrated in the upper Ohio River valley, flourished from about 2,800 years ago to about 2,200 years ago but had little influence on older Archaic life modes on the Great Plains. The Mississippian, concentrated in the lower Mississippi River and its tributaries, did not appear until about A.D. 900, long after the close of the Archaic period. The Hopewell, on the other hand, which arose about 2,200 years ago and extended through the Ohio and Mississippi valleys and along the Gulf Coast, had an important impact on the new cultural traditions emerging on the Great Plains.

In the Southwest the distinct cultural communities included the Hohokam, Mogollon, and Anasazi. Two of them, the Mogollon and Anasazi, also had influence over the emerging modes of life on the Great Plains. The Mogollons were a horticultural and collecting people whose culture was centered at Casas Grandes in northern Mexico but spread through the mountains of southeastern Arizona, southern New Mexico, and far western Texas. The Anasazis, often called the Basket Makers or the Cliff Dwellers of Mesa Verde, lived in what today is known as the "four corners" region, where Arizona, New Mexico, Utah, and Colorado come together.

During the peak of their development, the Mogollon and Anasazi peoples had large populations and complex social organizations. They traded baskets, blankets, and pottery to people far beyond their territory. Along regularly used, overland routes traders carried goods to the Pacific and Gulf coasts, the Great Plains, and elsewhere in exchange for seashells, animal skins, and other items. The settled life, surplus goods, and community existence promoted significant advances in nonmaterial culture. Religion, dance, and music became well developed, for example, and the artistic creativity and folklore of the people assumed new and important roles in their lives.

The Archaic period at the Lubbock Lake site, had lasted for some six thousand years. During that time humans had refined their hunter-gatherer lifestyle, shifted and diversified their economies, and produced a more sophisticated tool kit. They had successfully adjusted their economic strategies to a period of prolonged drought—the Altithermal. They had moved about in small hunting bands over a large territory, maintained contact with kin groups in neighboring bands for social needs and trade purposes, and began using dogs as beasts of burden. At the close of the Archaic period, new cultural groups began to appear on the Llano Estacado, just as they had appeared in the East and Southwest. The initial groups on the Llano often looked to the Mogollons and Anasazis, whose influence can be seen in the Modern period, beginning with some of the earliest new groups to emerge on the Llano.

★

Chapter 4

The Modern Period

Surviving Great Change

AT THE LUBBOCK LAKE SITE and through much of the upper Brazos River system, the Modern period extends from about two thousand years ago to the late nineteenth century. It includes what some scholars investigating archaeological sites on the Llano Estacado often call the Ceramic, the Protohistoric, and the Historic subperiods, but other names and conflicting dates and times have been used.

Briefly stated, the Ceramic period, the earliest of the three, extends from about two thousand years ago to about A.D. 1450. The period saw the first extensive use of pottery on the Llano Estacado. Probably, then, a more settled lifestyle developed, at least in some places. The bow and arrow, originally developed by hunters in the Pacific Northwest, now came into use on the Great Plains. At the Lubbock Lake site, archaeologists have found a Ceramic-era hearth, modern bison remains, a corner-notched Scallorn arrow point, lithic tools and flakes, bone beds, and debris that indicate the place was a camping spot.

The Protohistoric period runs from about 1450 to about 1650. A projectile point called the Garza, a small triangular point with a basal notch, represents a significant defining trait of the period. Some, but not much, pottery is present, as people began again to use, or perhaps did not stop using, skin containers and to once again adopt a mobile lifestyle. They

The Three Sisters

Corn, beans, and squash represent the "three sisters" of early Native American horticulture. Indian people of Mexico, Guatemala, and Honduras (Mesoamerica) developed some of the first strains of the three crops, and eventually others spread the foods north and south from Central America.

Through thousands of years Indian people experimented with food varieties, crossing plants to create new strains and creating hybrids that could prosper in more northern climates. They developed corn, for example, from a small-cob grass called teosinte. Later people crossed it with Chapalote, a small-cob popcorn, to produce the real beginnings of corn. As time passed, people developed new types and strains, corn with a larger number of rows, for example, and corn resistant to drought or cold, and corn that produced better flour. The first corn varieties may have entered the Southwest about three thousand years ago.

Cucurbits (the squash and pumpkin group) came from South America and Mexico about five thousand years ago, and people of the Central Andes first domesticated beans about four thousand years ago. People carried beans and squash with bottle gourds, which Mesoamericans had cultivated approximately 7,000 years ago, into the Southwest about 2,500 years ago.

Corn, squash, and beans provided enormous benefits to people who cultivated the crops. To prosper and mature, the three plants required approximately the same combination of moisture, growing season, and summer heat. Corn removed nitrogen from the soil, but beans restored it, and squash could be grown in the same field with corn. And, importantly, among them the versatile three sisters provided a balanced, nutritious diet. Corn provided protein. Beans provided protein and amino acids, especially lysine, that aid the digestion of corn. Squash, which may not have the same high nutritional value, nonetheless could be dried and stored, and squash gourds could be used as containers.

---

traveled in cyclical patterns as they followed the bison herds. As the era opened, people who spoke Athapaskan, particularly ancestral Apache groups, climbed onto the Llano Estacado and pitched their camps at places like Blackwater Draw and the Lubbock Lake site. Near its end, the Protohistoric witnessed the appearance of the first Europeans, most of whom were Spaniard explorers.

The Historic period, from about 1650 to about 1875, represents a time of increasing Euro-American presence at the Lubbock Lake site. During the period, different Indian groups came and went, displacing others on the Llano Estacado and then seeing themselves displaced. Horses, cattle, hogs, sheep and goats, and other European animals arrived to compete for grazing land and living space with bison, pronghorns, deer, and other animals of the region. People from a European background brought with them new plants, weeds, and grasses that competed with native species for growing space and ultimately survival. Europeans also brought disease that had a negative impact on both human and animal life. Clearly, then, the Historic era at the Lubbock Lake site and on the Llano Estacado, including the upper Brazos River region, represents a time of great and rapid change.

Not everything changed at once. Although the Archaic period had come to an end, some Archaic life patterns continued deep into the Ceramic, Protohistoric, and Historic subperiods. In the upper Brazos River region some of the earliest new cultural lifeways are associated with the Ceramic period.

On the Texas High Plains, Ceramic-age sites, which are numerous, tended to be small. They were also scattered and, because they often reveal little in the way of architecture, not much is known about house forms. Moreover, people occupied them for short periods of time.

Peoples of the Southwest, especially the Anasazi and Mogollon, and peoples of the Eastern Woodlands, particularly the general Woodland and Hopewell peoples influenced Ceramic-period groups. On the northern Llano Estacado, for example, the Ceramic-period people utilized corn and adopted the bow and arrow, both either obtained or copied from the Southwest, particularly from the Anasazis. They became more sedentary so that they might tend to their small fields of corn and perhaps beans and squash, collectively called the "three sisters," plus such older crops as amaranth (pigweed) and marsh elder. They also began to make two types of pottery. Some was cord-marked, possibly an Eastern Woodlands adaptation, and some was smooth and brown, implying a Southwestern adaptation—probably a style related to eastern Mogollon called Alma Plain.

In adopting pottery making, the people still retained many older customs. They hunted, of course, but now using bows and arrows, and they continued to forage for wild plants. Social and cultural life in many cases continued to represent Archaic ways. In fact, at some of the sites the presence of pottery seems to be the only manifestation of new modes of life, suggesting only a small change in the Archaic complex.

Some early Ceramic sites have been identified along the Canadian and

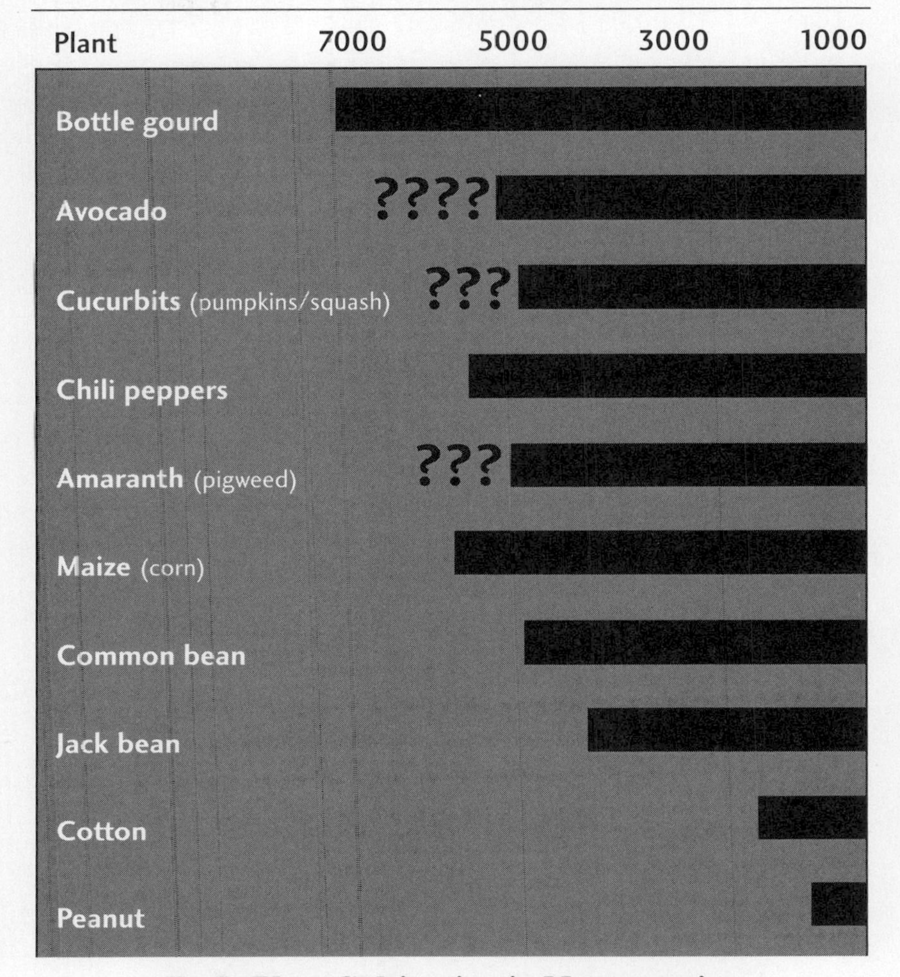

Early Plant Cultivation in Mesoamerica
Modified from Robert F. Spencer, et al., *The Native Americans: Ethnology and Backgrounds of the North American Indians,* 2nd ed. (New York: Harper & Row, Publishers, 1977), 23.

Red rivers in the Texas Panhandle and east of the Pecos River in southeastern New Mexico along the Llano Estacado's Mescalero Escarpment. The more northerly Panhandle sites take the name Lake Creek. Dated tentatively from about A.D. 200 to A.D. 900, the sites exist on the upper edge of the Llano Estacado along terraces of tributaries to the Canadian River. Lake Creek people maintained an economy based on foraging over a wide area and for a wide assortment of foods. They probably did not rely much on bison hunting, but they interacted with people of the Southwest, particularly the Jornada Mogollon, for brown, smooth shards known as Alma Plain have been found. Eastern influences are present as well, for cord-marked ceramics are present. Ceramics from foreign sources suggest the existence of distant trading partners.

In fact, Lake Creek people were traders. They traded with Woodland peoples to the east and Southwestern desert peoples to the west. Lake Creek groups mined the Alibates quarries and used the flint for tools and for trade. Their extensive foraging probably took them to neighboring villages, and their unique flint, by way of long-range traders, moved even farther, sometimes great distances. Alibates flint has been found at archaeological sites all across the United States and even deep into Mexico. In turn, perhaps, foreign traders visited their villages. The presence of such stones as obsidian and Edwards Formation chert also suggest trade had reached across long distances.

A similar culture, the Palo Duro, existed some thirty miles to the south along the upper waters of the Red River. It dates from A.D. 300 to A.D. 900. As at Lake Creek, some cord-marked pottery exists at the Palo Duro sites amid larger amounts of the smooth-surfaced, brown, Southwestern-derived Alma Plain. Also like their neighbors along the Canadian, Palo Duro people maintained a subsistence economy that centered on foraging with little reliance on bison. Archaeologists have found earthen and rock-lined fire pits and pit structures.

A later culture, one dating from 950 to 1100, existed along the southwestern edge of the Llano Estacado. Here Mogollon interaction with people on the High Plains shows greater influence than in the Texas Panhandle. Known as the Querecho, it may represent an indigenous development, but some scholars insist that the people were an eastern Jornada group that over time became independent of other Mogollons. Archaeologists have found open campsites along the Mescalaro Escarpment, but they have identified no structures, except perhaps a few small, rectangular pit houses. Small, corner-notched projectile points suggest Querecho people used bows and arrows to hunt small to medium-size game: rabbits, pronghorns, and deer, for example. The people used metates and manos to grind acorns and mesquite beans. We do not know if the Querecho people cultivated corn.

Alibates Flint

Alibates flint is an agatized dolomite—that is, in this instance, a limestone rich in magnesium carbonate. Characterized by its distinctive coloring and banded patterns, Alibates flint often contains maroon and white colors, but blue, brown, red, or yellow may be present.

Alibates flint quarries exist in few places. But along the Canadian River near modern Fritch in the Texas Panhandle a whole series of quarries can be found, a few of them about three or more feet deep and ten feet in diameter. The outcrops, many of which are located along canyon rims, cover an area of about ten square miles. Some of the quarries today are part of the Alibates Flint Quarries National Monument, some 1,079 acres containing hundreds of small quarries and several sites that had been Indian villages.

Indian people from the time that humans first occupied the Southern High Plains visited the Canadian River quarries. They dug out the flint or gathered it from gravel downstream from the outcrops. They used the highly workable flint for the manufacture of tools and weapons, such as knife blades and projectile points. Paleoamerican groups, Archaic people, and other Native Americans through to modern times all sought Alibates flint. They used the flint themselves or traded it to neighbors, who in turn traded it to more distant people. During the Ceramic period (two thousand years ago to A.D. 1450), sedentary village people in the Texas Panhandle traded Alibates flint for Puebloan ceramics from the Rio Grande valley in New Mexico.

Archaeologists have found caches of Alibates flint—and sometimes small Alibates quarries—over wide areas of the Southern Plains, including the Lubbock Lake site.

At the Lubbock Lake site, for the same period, little evidence has as yet been found for maize (corn) horticulture. But the presence of Chupadero black-on-white pottery suggests an interaction with Southwestern cultures, where people grew corn. Archaeologists have found Alibates flint and Edwards Formation chert in Ceramic-period deposits, suggesting a mobile society, extensive trade, or both. Seeds of wild plants exist, but no seeds relating to early horticultural products, such as amaranth, corn, beans, or squash, have been reported.

After 1200, Ceramic-age sites on the Llano Estacado increased. Perhaps the most famous are the more than one hundred sites associated with the Antelope Creek culture. Located along the Canadian and North Canadian rivers in Texas and Oklahoma, they date from about 1200 to 1500. Antelope Creek is a local development that probably grew from people of the Lake Creek culture. The Alibates flint quarries seem to be a key to the Antelope Creek existence, for the sites suggest abundant trade contacts to the Southwest and to the Central Plains, particularly the Republican River in present Kansas and Nebraska. Indeed, Antelope Creek may have represented something of a distribution center for Alibates.

The people of Antelope Creek were horticulturists who also hunted and foraged. They raised crops of corn, squash, pumpkins, and beans; hunted bison, pronghorns, and deer; and gathered plant foods, including fruits, berries, tubers, and nuts. They fashioned implements and weapons from bison bones, including hoes, digging sticks, awls, wedges, rasps, and knives. They manufactured drills, scrapers, and a variety of other tools. They lived in multiroom, stone-slab homes reminiscent of but not related to Southwestern pueblo architectural styles.

Antelope Creek populations increased over the years, but then rather suddenly they declined. Drought and changing socioeconomic conditions, including the appearance of such Athapaskan groups as the Apaches, led to their dispersal. Perhaps they merged with Garza people who lived on the edge of the Llano Estacado to their southeast or moved northeastward to join ancestral Pawnees living along the upper Republican and Arkansas rivers.

Despite changes in life patterns as reflected among Antelope Creek, Querecho, Palo Duro, and Lake Creek peoples, many socioeconomic groups on the Llano Estacado continued Archaic hunter-gatherer traditions. In particular, most Indian people who used Blackwater Draw or the Lubbock Lake site continued essentially Archaic-age life patterns. Evidence of permanent dwellings has not been found at the Lubbock Lake site, nor has evidence of corn-based horticulture. But archaeological activity at the site, directed by Eileen Johnson of Texas Tech University, is ongoing. Perhaps the investigations will reveal new information about the people and their lifeways on the Texas High Plains.

Nonetheless, all groups associated with the Llano Estacado during the Ceramic period underwent change and transformation. Identifiable cultural groups appeared, prospered for a time, and as a result of environmental change or sociopolitical pressures declined. Indian people associated with most of the sites maintained a sedentary village life with an economy based on horticulture and wide-spectrum foraging. Horticulture, however, nearly always remained marginal and associated only with sites in valleys or river drainage systems where water was readily available. The various peoples used the Llano uplands for hunting and foraging, subsistence activities that were intense enough to slow the development of full-scale horticulture.

The Ceramic period did not endure very long. By 1450 new developments had begun to impact people of the Llano Estacado. In one development, strong groups of Athapaskan-speaking people, dominated by ancestral Apaches, migrated southward from their northern homelands. Their push toward the south, which took paths through the Great Basin or over the Great Plains along the front range of the Rocky Mountains, began sometime after 1200. By 1450 they had reached plains village sites along the upper Plate, Republican, and Arkansas rivers. They challenged the Antelope Creek people after 1450, contributing to the demise of sedentary village life in the Texas Panhandle.

Drought was a second development. After 1435 a series of prolonged droughts hit the Southern Plains. Spanning a combined total of thirty years, the long dry spells contributed to successive crop failures. They also represented important reasons for causing people such as those at Antelope Creek to abandon their horticultural villages and disperse. Elsewhere over the Great Plains similar droughts combined with the arrival of Athapaskan-speaking people from the north to force once-flourishing village people in the upper Republican River to move on.

As the Ceramic subperiod ended on the Texas High Plains, the Protohistoric period, circa 1450–1650, began. Socioeconomic groups dependent on bison hunting returned to the Llano Estacado.

One of the new groups was Tierra Blanca. People of this mobile, bison-hunting society were apparently ancestral Apaches who had for decades been moving south from northern homelands. In the published reports, the Tierra Blanca society is not well defined. The people probably were related to the Querechos, about whom Francisco Vásquez de Coronado and his chroniclers of the 1540s wrote. The Querechos have been identified as ancestral Lipan Apaches.

The people of Tierra Blanca pushed out the Antelope Creek groups and established themselves in the region, but they did not practice horticulture. Some ceramics of a Southwestern variety have been associated with Tierra Blanca, granted, but the people were primarily hunter-gatherers who used small triangular notched and unnotched points. They did not occupy the abandoned stone-based houses of the Antelope Creek people but lived in tipis, used skin containers, and employed dogs to help carry their possessions. They hunted over wide areas of the Texas High Plains concentrating on bison, using bows and arrows, and reestablishing a mobile life in the region.

Coronado in 1541 met and talked with Querechos in the vicinity of Frio Draw and Tierra Blanca Creek in modern Parmer County. Elsewhere amid large herds of bison he saw other Querecho groups—but did not always converse with them—in the upper Brazos River region. Coronado and his chroniclers indicated that large dogs pulled travois loaded with Querecho camp goods and personal possessions. He also reported that the people sometimes ate raw bison meat, drank animal blood when water was not available, wore clothing of tanned skins, and moved about from one campsite to another. The Querechos, who were superb hunters, pursued bison on foot using bows and arrows. They alternately traded with and raided such Pueblo towns as Cicuye, or Pecos, located southeast of modern Santa Fe and nearest of all pueblos to the Llano Estacado. The trade included plant foods and village wares from Pueblo people for meat, hides, robes, and other bison products from the Querechos.

The Garza society, circa 1450–1650, represented a second new group. Centered, perhaps, among a whole series of small villages in Blanco Canyon but stretching westward through the Llano Estacado to the Pecos, it included the Lubbock Lake site, Blackwater Draw, and other sites in the upper Brazos River region. Garza people were hunter-gatherers who relied on bison hunting, food collecting, and extensive trade with both Puebloan peoples to the west and Caddoan speakers to the east.

At the Lubbock Lake site, archaeologists have identified camping areas and processing stations associated with Garza people. Their excavations have found no structures, but they have revealed some living spaces, smooth areas within a circle of holes that suggest tents or tipis. They have discovered small triangular, notched points, snub-nosed scrapers, and stone drills and knives. They have located some bone implements, such as awls, which formed an important part of the Garza tool kit. Also at the Lubbock Lake site, Edwards Formation chert dominated the stone artifacts, but some Alibates flint was present.

The identity of the Garza population remains uncertain. Nearly all scholars agree that Garza people represented some of the scattered groups that Coronado and his chroniclers called the Teyas. They also seem to be in agreement that the Teyas were a tattooed people who maintained several important hunting and trading villages along the eastern caprock and in Blanco Canyon. And they seem to agree that the Teyas traveled to Pueblo villages, especially Cicuye, for trade.

But here the agreement ends. Most anthropologists, archaeologists, and historians argue for an eastern background for the Teyas, meaning that Garza groups were Caddoan speakers, possibly ancestral Wichitas, who regularly came to West Texas to hunt bison and conduct trade. They suggest that some of the dispersed Antelope Creek and related folk of the plains had joined Garza people.

A few scholars maintain that the Garza complex represents an ancestral Apache group, one somewhat different from the Querecho/Lipan Apache ancestry. In such an interpretation, Teyas become Athapaskan speaking, ancestral Mescalero Apaches.

Still other authorities claim that the Teyas were ancestral Jumanos, a people whose territorial base later covered the Edwards Plateau between the lower Pecos River in Texas and the Colorado River. In this view the Teyas spoke a Uto-Aztecan language, Tompiro, and derived important associations with Tompiro people of the Salinas Pueblos in New Mexico. Their ancestors, so the argument goes, had drifted eastward, become bison hunters, and moved onto the Llano Estacado.

The Teyas in Coronado's view were friendly people. Like the Querechos, they hunted bison on foot using bows and arrows. They colored their eyes and chins with dark, distinctive tattoos that they formed by filling cuts with charcoal. In Blanco Canyon they collected a plentiful supply of wild plants, fruits, and nuts. If Coronado's chroniclers are correct, the Teyas cultivated some beans. They lived in scattered villages or rancherias through much of the lush, well-watered canyon, and they climbed out of the wide gulf to hunt bison. Like the Querechos, the Teyas employed the plains sign language for communication with foreign groups.

The Tierra Blanca/Querechos and Garza/Teyas dominated the Llano Estacado and the upper Brazos River system during the Protohistoric period, but they were not the only groups in the region. Caddoan-speaking ancestral Wichitas, for example—whether or not they were Coronado's "Teyas"—came to the plains to hunt and trade. They favored the upper Red River drainage area.

Village-dwelling ancestral Jumanos—different from, but perhaps distant kin to, Coronado's Teyas—came to the Texas High Plains. They resided in permanent towns at La Junta, located on the Rio Grande near modern Presidio, Texas, and visited the southern Llano to hunt and trade. They also traded with Pueblo people in southern New Mexico. Moreover, their kinsmen, the mobile bison-hunting Jumanos, often traveled to the La Junta area to secure corn, squash, beans, and other products from the horticultural villages.

But what became of the Ceramic-aged Querecho complex along the Mescalero Escarpment? What happened to these people of the Llano's edge after their society declined around 1100? If the Jumanos of La Junta are descendents of a Jornada Mogollon population indigenous to the region, perhaps dispersed Querecho people, a Jornada Mogollon group, formed part of the La Junta ancestry.

Whatever the case, additional changes came after Europeans arrived. On the southern plains Álvar Núñez Cabeza de Vaca and three companions in the 1530s skirted the region through Texas and northern Mexico. They passed through the La Junta villages, but they did not reach the Llano Estacado.

A few years later, however, in 1541, Francisco Vásquez de Coronado came to the Llano Estacado. He led a force of some 1,800 soldiers, Mexicans, Indians, servants, other auxiliaries, and even some recent captives from the Tiguex pueblos, near modern Albuquerque, New Mexico. He probably climbed onto the Llano somewhere between modern San Jon and Ragland, the historic Arroyo del Puerto area, in present-day Quay County, New Mexico.

Coronado's route across the Texas High Plains remains another of those wonderful mysteries with which scholars grapple. Recently, John Miller Morris, a geographer at the University of Texas at San Antonio, argued impressively for a route that took Coronado and his force southeastward, across Frio Draw at its far upper reaches first and then Running Water Draw near the modern New Mexico–Texas boundary. Continuing

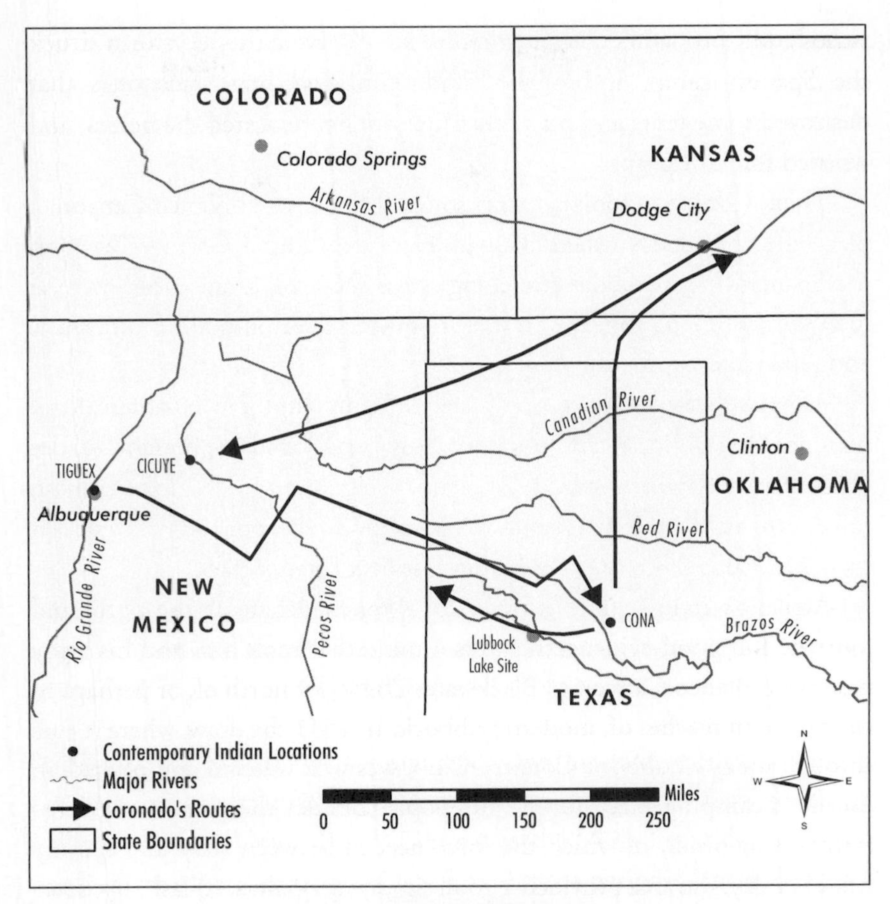

Coronado's Routes

in a southeasterly direction, Morris claims, they struck Blackwater Draw at a point deep in the heart of the Llano Estacado's most impressive flatness and from there turned northeast. They again crossed Running Water Draw, this time between modern Edmonson and Plainview, Texas, and several days later they reached the Llano's eastern escarpment somewhere in or near modern Briscoe County, the general area around and below lower Tule Canyon.

Here Coronado rested a few days amid camps of Teyas. The Native Americans provided an assortment of food for Coronado's large force, which probably had become tired of an endless round of bison meat. They also provided information on securing available food sources to feed 1,800 people twice a day and the direction to Quivira, supposedly a land of great

tohistoric period ended—around 1650. The increasing contact with additional European trade goods and the concomitant development of a horse complex among Indian people also marks the beginning of the Historic period in the upper Brazos River region. The Europeans coming to the Texas High Plains included explorers, missionaries, traders, adventurers, and others.

In passing through the Texas High Plains, European and Indian people used, basically, one of three routes. A northern route followed the Canadian River. A central route, presuming one started in New Mexico, took sojourners from Cicuye, Pecos pueblo, southeastward to Tierra Blanca Creek and the upper Frio Draw and from there as directly as possible to the Colorado River and Concho River country. Some may have followed Blackwater or Yellow House draws across the Llano before moving south along the eastern caprock, but others probably crossed along a course that took them along Seminole Draw. A southern route circled around the southern reaches of the Llano Estacado, connecting the Edwards Plateau country with the Rio Grande pueblos by crossing the lower Pecos River. The availability of water governed route selection, whether heading east or west.

How often Spaniards, using one of the central routes, came through the Lubbock Lake site remains unknown. Between 1629 and 1654 Spanish missionaries and traders were busy proselytizing to Jumanos and trying to collect freshwater pearls from the Concho River in the modern Ballinger–San Angelo region. Probably some of the Spaniards, in their coming and going during the brief period, followed either the Yellow House or the Blackwater draw—parts of the central routes. If so, perhaps some of them traveled by and stopped at the Lubbock Lake site.

More important, however, we know that Spaniards visited the Texas High Plains. They explored, traded, and preached. During the seventeenth century they renamed practically every major river, stream, lake, canyon, or other landmark in the region. Consider, for example, *Punta de Agua* (Point of Water/Lubbock Lake); *Casas Amarillas* (Yellow Houses), *Los Brazos de Dios* (The Arms of God/Brazos River), *Agua Corriente* (Running Water), *Atascosa* (boggy or muddy) Spring, *Laguna Plata* (Silver Lake), and *Cañón de Rescate* (Ransom or Traders Canyon). Consider also

Apaches

Ancestors of today's Apache people arrived in the Southwest from present-day Canada probably around 1400. Athapaskan speakers, they are related to Navajos, northern Denes, Beavers, and other groups in northwestern Canada. They filtered into the mountains and plains surrounding lands of the Pueblo people in the upper Rio Grande in New Mexico, Colorado, and Texas and spread out from this area. Francisco Vásquez de Coronado in 1541 met ancestral Apaches in Texas while searching for his Gran Quivira.

In a broad sense the Apaches split into two large divisions: eastern and western. The Western Apaches came to include such modern tribal groups as Cibecue, White Mountain, San Carlos, and Northern and Southern Tonto. Historically they ranged through Arizona, western New Mexico, and northern Mexico. They were a people who preferred cool, timbered locations in the southwestern uplands and mountains. Some people might include the Chiricahuas among the Western Apaches. The Eastern Apaches came to include the Mescaleros, Jicarillas, Lipans, and Plains (Kiowa-Apaches). Historically they ranged through the mountains and plains east of the Rio Grande. After they acquired horses in the late seventeenth century, they became mobile, equestrian bison hunters who traded products from bison (hides, skins, meat) to Pueblo villagers for vegetable foods.

Eastern Apache groups alternately warred against and traded with Pueblos and Spaniards. Comanches, who showed up in New Mexico around 1700, ultimately drove the Eastern Apache groups from the Texas High Plains.

In most Apache groups, gender roles were clearly defined. Women collected and prepared food, took charge of the home, cared for the children, and prepared animal hides and skins. Men hunted, raided, and waged war; prepared weapons; and among the eastern tribal groups took charge of the horses and equipment. Most Apache groups were matrilineal (tracing descent through the maternal line) and matrilocal (husbands and children associated with the kin of the wife and mother).

Palo Duro (hard or dry wood) Canyon, *Tule* (reeds) Canyon, *Colorado* (red) River, *Tierra Blanca* (white earth) Creek, *Tecovas* (covering) Spring, and *Blanco* (white) Canyon.

The Pueblo Revolt in 1680 brought new people to the Lubbock Lake site. In it, Pueblo people rose up against Spanish authority in the upper Rio Grande country, killing many of the Europeans and forcing others to retreat downriver. The Spaniards regrouped at modern El Paso and twelve years later began the "reconquest" of New Mexico.

During the Spanish absence, the village-dwelling Pueblos allowed

Apaches, Utes, and others to collect horses that Spaniards had left behind. Over a two- or three-year period Indian people removed some ten thousand animals, nearly sweeping the place clean of horses. Quickly, the eastern Apaches became expert horsemen and on the Southern Plains adapted the animals to their mobile, bison-hunting way of life.

Apaches, linguistic relatives of the Navajos, divided into several branches and, as today, went by a variety of names. Some of the first Europeans to meet and describe Apaches no doubt used different names to identify essentially the same group. Three of the eastern groups, the Lipans, Jicarillas, and Mescaleros, ranged over parts of eastern New Mexico, West Texas, and the Southern Plains in general. For generations their people traded at Cicuye pueblo, mainly Lipans who have been called Quecheros, Faraones, Ypandis, and other names; Taos pueblo, mainly Jicarillas; and Picuris pueblo, probably Mescaleros.

But the Pueblo-Athapaskan relationship was always in flux. Sometimes Apache people spent winter months near Pueblo communities. Sometimes, particularly after drought years had left Pueblo corn harvests short, the Apaches and Pueblos fought over trade or other issues. Sometimes the Apaches and Pueblos united against the Spaniards. The acquisition of horses and the adoption of an equestrian way of life after the Pueblo Revolt clearly gave the Apaches an important advantage over the Pueblos. Indeed, Apache groups soon dominated a huge region in the Southwest east of the Rio Grande.

A second development affecting the Llano Estacado was the arrival of Comanches. The Shoshonean-speaking Comanches, linguistic relatives of the Utes and Shoshonis, left their northern homelands and moved south through the Great Basin. Before 1700 they were fighting Utes, their former kinsmen, and were already mounted on horseback. In 1706, if not before, they showed up to trade at the Taos pueblo. Attracted by a warmer climate, a large supply of horses, and enormous bison herds on the central and southern Great Plains, they invaded Apache country.

The Comanches became powerful and aggressive. In the first few decades of the eighteenth century, they fought Caddoes of eastern Texas and Louisiana and their allies, which now included French nationals from the Mississippi valley, to the east; Pueblos and Spanish soldiers in New Mexico; Utes to the northwest; and Apaches on the Southern Plains. Only

Comanches

Shoshonean-speaking Comanches, perhaps under pressure from other peoples, began moving south from Wyoming through the Great Basin in the seventeenth century. After 1700 they were in New Mexico, mounted on horses, hunting bison on the Great Plains, and adopting a Plains Indian lifestyle. Aggressive and powerful, they pushed the Apaches from the western plains of Texas, fought with the Spaniards, alternately traded with and raided the Pueblo groups of New Mexico, and expanded their range of territory eastward and southward through Texas.

In the eighteenth and nineteenth centuries the Comanches comprised several culturally similar but politically independent tribal divisions. The divisions came and went, combined and split, and rose and fell depending on political concerns, economic resources, and tribal populations. In the mid-nineteenth century the Comanches counted six divisions: Yamparika, Kotsoteka, Tenewa, Penateka, Nokoni, and Kwahada. Within the divisions the people often organized in mobile bands of extended families that were linked to the larger units by kinship, sodality, and political ties.

The Comanches were bison hunters mainly, but in some ways they are better understood as horse pastoralists. They traded bison products, including hides, skins, and meat, with their neighbors both to the east and to the west for vegetable fare and, later, manufactured goods. Men hunted. They also raided their neighbors, including communities in northern Mexico, for horses, human captives, and material goods. They fashioned weapons and protected the village. Superbly skilled horsemen, they also trained and guarded their large horse herds. Women took charge of the tipis, gathered wild plants and other foods (plums, grapes, nuts, berries, onions, and wild potatoes, for example), prepared the meals, fashioned the clothing, and looked after the children.

For a century or more the Comanches were the dominant power on the Southern High Plains, ranging across a territory in Texas, New Mexico, and Oklahoma that became known as *Comancheria.* They were, as many authorities have noted, lords of the Southern Plains.

---

the Apaches were formidable opponents, but they, too, eventually gave way.

In most ways, the Comanches became typical Plains Indians. They were equestrian bison hunters who lived in tipis, kept dogs, and used travois. They admired skill in hunting and warfare and honored such success by creating large, feathered headdresses. They maintained warrior societies, utilized the plains sign language, and adopted the plains vision quest—a nearly universal plains religious activity in which individuals sought dreams and personal revelations or "visions" to guide one's future course. Men hunted, defended encampments, and dominated decision making. Women collected vegetables, cooked, tanned bison hides, made clothing, and tended to household duties.

Politically, Comanche people did not form a single tribal unit. Rather, they divided into several independent divisions. Nor were the divisions static. They arose, merged with other branches, split into new groups, disappeared. Rarely did the separate groups all come together, but Comanche people respected their kinsmen and often communicated and cooperated with people from neighboring divisions. As a mobile, equestrian people, they organized, socioeconomically and politically, into hunting bands led by warriors and respected elders. At their peak of power and influence in the early nineteenth century, the Comanches dominated the country below the Arkansas River all the way south to the Balcones Escarpment, west of the Texas Cross Timbers, and east of the upper Rio Grande—a region called *Comancheria.*

In western Texas, Comanche divisions visited the canyons and brakes of the Llano Estacado. Indeed, they often spent winter months in such canyons as Palo Duro, Tule, Quitaque, Blanco, and Yellow House. In the nineteenth century, streams in the canyons still ran with water. Springs issuing from the Ogallala Aquifer were abundant, some quite large.

At the Lubbock Lake site, active springs fed a ten-acre lake. Except in extreme drought years, enough water flowed from the lake to maintain a stream of clear, cold water one foot deep and twelve feet wide through much of the lower Yellow House Canyon. In present-day Mackenzie Park, a long lake once existed below the juncture of the Yellow House stream and the water coming down Blackwater Draw. Farther downstream, Buffalo Springs added water, and in modern Ransom Canyon, which forms a small part of the larger Yellow House Canyon, springs along Pig Squeal Creek and the east wall contributed to the flow.

Plants and grasses in Yellow House Canyon and at Lubbock Lake were abundant for both bison and Indian groups. For bison, tall prairie grasses existed near the water and shorter, matted grasses, such as buffalo and mesquite grass, spread above the valley floor and onto the Llano Estacado. For the Comanches, hackberry trees with their little red berries grew in the breaks, and mesquite trees provided bean pods. Wild grapes, plums, mulberries, sunflowers, wild onions, and other plant foods added to the larder.

Lubbock Lake in the late 1950s. Courtesy Southwest Collection: SWCPC (W) E3.22.jpg.

They also utilized various cactus fruits, particularly those of the cholla and prickly pear.

Despite the rich grasslands, the Comanches were increasingly concerned about feed for their great horse herds. Comanche families possessed many horses with some individuals owning fifty or more animals. Horses fed on the same grass that bison consumed. Consequently, as the horse herds increased in size, the need to find adequate forage for the animals governed the selection of camping sites. As time passed the Comanches began to look more like horse pastoralists driving their animals to better pastures than like mobile bison hunters seeking a good stand.

Furthermore, the Comanches came under increasing pressure from Europeans. Spaniards from Santa Fe and San Antonio had been battling Comanches for a century or more, trading with them and negotiating treaties with them. In one such accord, the Pecos Treaty of 1786, western Comanches—those associated with the Llano Estacado—and Spanish authorities of New Mexico agreed to peace between them but to war against the Apaches. The far-reaching treaty allowed Spanish explorers to seek a road between Santa Fe and San Antonio. It also permitted *comancheros,* mainly Pueblo and Hispanic traders from upper New Mexico, and *ciboleros,* bison hunters, to enter the plains with their ox-drawn, two-wheeled carts to trade with plains people and to hunt bison.

Soon, comancheros were all over the Texas High Plains. They held annual trade fairs in Quitaque, Blanco, and Yellow House canyons. They established semipermanent camps in Casas Amarillas and south of modern Post near Muchaque Peak. They visited Lubbock Lake. The journey from New Mexico, which took at least four weeks, carried some traders along the Canadian River, the old northern route through the Llano Estacado. Most comancheros used the central route, hauling their two-wheeled carts down one of the upper arms of the Brazos River. Trading activities sometimes lasted a month or two, after which the comancheros started home, another journey of four or five weeks—unless they were driving cattle they had acquired in the trade, a circumstance that lengthened the trip.

Then came the Americans, mainly descendants of the British colonists, who arrived in large numbers in the West and Southwest after the turn of the nineteenth century. In Texas, traders and adventurers led the way. Cattlemen, farmers, and townspeople were not far behind. The Americans pushed beyond the Mississippi River, confronted Spaniards first and, after 1821, Mexicans; fighting for rights to the land, they took over Texas. Just as aggressive and as expansive as the Comanches, the Americans represented a new challenge to Comancheria. American traders replaced Frenchmen along the major plains waterways, and American soldiers were not far behind. As Americans advanced, Indian people resisted, and trouble followed. Subsequent warfare—coupled with a long series of disease pandemics, especially smallpox, that stretched back to the seventeenth century—resulted in population decline for natives all over the West. Reports of food shortages and hungry Indian people mounted, especially in the 1830s and afterward.

Also in the 1830s American fur traders reached the Texas High Plains. In 1832 Albert Pike, a tall, slim, combination lawyer, poet, scholar, and diplomat, and Bill Williams, a tall, raw-boned, red-haired fur trapper described as a superb hunter, joined a group of trappers in Santa Fe and headed east. The men planned to trap beavers in some of the head streams of rivers in the Rolling Plains.

The group followed Blackwater Draw southeastward. Thus, it came through the site of modern Lubbock, reached Yellow House Canyon in Mackenzie Park, and continued down the Yellow House. At the mouth of the canyon it turned north across the Salt Fork of the Brazos River and headed for the Pease, Washita, and Red rivers.

Pike wrote about the trip. He described comancheros and their trails and commented on Comanches and their camps. He noted the topography of the country and described the region's plants and animals. He indicated that his group of approximately forty-five hunters was headed first to *Cañón de Rescate* (Ransom or Traders Canyon), southeast of present-day Lubbock, and then to the "beaver meadows," as he called them.

Pike wrote in some detail about Blackwater Draw. In the fall of 1832, he noted, the upper draw was dry, but the men could get water by digging holes in the creek bed. Others, he suggested, had done it before them. Farther down the draw they found springs that provided cool, clear water, abandoned Indian camps with lodge poles, and piles of bison bones. They saw pronghorns and scattered bison bulls. There were no trees, except a few scrub oak bushes. They made cooking fires with horse dung and dry, twisted grass. A few days later they found a few trees, the first they had seen since leaving the Pecos River. Still farther down the draw, probably near the present-day Lubbock Country Club, Pike noted that hackberry, mesquite, and cottonwood trees could be found. He also wrote that small ponds of water were available in the draw and, upon reaching modern Mackenzie Park, a shallow stream twenty feet wide ran with fresh water, which others have dubbed the "long lake."

Pike's party encountered three Comanche camps. The smallest one held twenty lodges and the largest, located in Yellow House Canyon near modern Buffalo Springs Lake, contained fifty lodges. Together, the camps contained, Pike estimated, at least five thousand horses. The people of the smaller village, Pike reported, were poor and shabbily dressed with no blankets, no bison robes, and no meat. People in the larger camp were better dressed and better provisioned. Besides meat, they had grapes, cactus fruits, hackberries, and nuts.

The Pike-Williams party represents one of the first Anglo American groups to cross the Llano Estacado. Pike's observations in 1832 about large Indian horse herds, the poor material circumstances of some Indian bands, and the busy comanchero trade reflect changing conditions on the Texas High Plains. Bison numbers already were in serious decline. Pike reported seeing less than ten scattered bison bulls during the two-week crossing of the Llano. Also, with Indian people pasturing large horse herds in protective canyons of the Llano, bison had fewer places to spend the winter and more competition for grass luxuriant enough to sustain large herds. Accounts of Indian food shortages, such as that reported by Pike, were becoming common.

Clearly, changes of many kinds had come with increasing rapidity in the Modern period. Apaches had evicted some of the village folk, such as those at Antelope Creek, and Comanches in turn had evicted Apaches. The mounting presence of white settlers, especially Spaniards and Americans, had meant additional alterations in native occupation of the Llano Estacado.

One of the dramatic changes is reflected at the Lubbock Lake site. Once the little lake watered thousands of bison and represented an important hunting ground for native people of the Llano Estacado. When Albert Pike visited Yellow House Canyon, the lake watered thousands of horses and served as an important camping site for native people with large horse herds.

Soon, that is, as soon as Americans pushed toward and onto the Llano Estacado, the Lubbock Lake site would water thousands of cattle. And, in terms of deep time, very quickly afterward it became a place where humans began digging through dirt.

★

Chapter 5

Reconstructing Deep Time

Digging Through Dirt

OVER THE PAST 170 years, human activity in Yellow House Canyon has changed and by changing has altered the Lubbock Lake site. Since the wet, rainy days of October 1832, when Albert Pike, Bill Williams, and their party of trappers traversed Blackwater Draw and Yellow House Canyon, shifting weather and climate patterns have also affected the site. Not all has changed, but by digging through soils and dirt in the canyon, scientists know that humans have occupied the Lubbock Lake site through at least a portion of deep time.

It is clear that Lubbock Lake and Yellow House Canyon have always attracted humans. Through the mid-nineteenth century, about 1835 to 1875, Comanches and their Kiowa allies, plus occasional Apache groups from New Mexico, hunted, camped, and grazed their horses at Lubbock Lake. Comancheros traded there, in Ransom Canyon, and at points farther down the Yellow House. Their roads, which sometimes stretched fourteen or fifteen horse trails wide, were broad and plainly visible. After the Civil War, American soldiers in pursuit of Indian warriors marched up the canyon and camped at Lubbock Lake.

One military officer, Lieutenant Colonel William R. Shafter, in 1875 described the area. A well-worn comanchero trail, he said, extended from the Lubbock Lake site about thirty miles almost directly east to Silver Falls in Blanco Canyon. The high tableland between Yellow House and Blanco contained many small round playa lakes that were filled with water for part of the year. The region was covered with luxuriant grass that supported immense herds of bison. Shafter correctly noted that the land could support thousands of cattle, even in years when the playas dried up, because plenty of water existed in Blanco and Yellow House canyons. He reported that wood—mesquite, cottonwood, and hackberry—was available. Corn, he thought, could be grown in Blanco Canyon without irrigation, except in unusually dry years.

Shafter's men were likewise observant. On the divide between Blanco and Yellow House canyons, they met nine comancheros with trade goods on pack mules. Shafter employed the comancheros as guides. One of his men reported that for several days in early August, southwest of the Lubbock Lake site, the military command rode through bison herds estimated in the hundreds of thousands. The animals, he noted, would move off to a safe distance on either side of the soldiers and turn to gaze at the intruders.

Not all was pleasant. Along some of the streams and at some water holes, such as at the Lubbock Lake site, mosquitoes troubled both men and horses. In fact, one of Shafter's officers remembered that mosquitoes in 1875 were terribly thick—ten thousand to the inch, he said. Camp fires and smudge pots did little good.

Also in the 1870s, Anglo bison hunters climbed onto the Llano Estacado. Having destroyed the central herds in Kansas, Nebraska, and eastern Colorado, the hunters, risking retaliation from Comanches and Cheyennes, moved into the Texas Panhandle in 1874. Indian warriors, as expected, struck back, and the United States Army in turn forced the Native Americans back onto their reservations in Indian Territory (Oklahoma). Three years later, the bison hunters established hunting camps all along the Llano Estacado's eastern escarpment, and a few moved up Yellow House Canyon to Casas Amarillas.

In response, hide buyers established Rath City. Located south of the Double Mountain Fork of the Brazos River in modern Stonewall County, the tiny outfitting post enjoyed a brief but colorful history. From a camp at little Silver Lake deep on the Texas High Plains, Comanches, both those who had not gone to the reservation and others who had later joined them, struck the hide town in May 1877, running off more than one hundred horses and mules—all the animals present.

Bison hunters, who were supplied horses by area ranchers, went in pursuit. For three months in the summer of 1877 the hunters operating on the Llano Estacado rode from water hole to water hole and back again looking for the Comanches, who had pulled out to sand hills in New Mexico, and their horses.

The hunters watched as climatic conditions impacted plant and animal life on the Llano Estacado. In the spring—May and June—rains had produced a rainbow of colors with the flowers in bloom, the playas full, and the streams running with fresh water. The tough, hardened bison hunters marveled at the beauty and abundance of the wildflowers and what they called the sweet perfume of the plains. The air was fresh and full of fragrant odors from verbena, black daisies, Indian paintbrush, buttercups, Tahoka daisies, and dozens of others. Yucca, with its light green stems and its tall, flowering stalk and white blossoms, added to the color. Above Cedar Lake in modern Dawson County they saw a few scattered stands of bison, and they met ciboleros from New Mexico. But below Cedar Lake they watched, awestruck, as hundreds of wild mustangs grazed over thousands of acres.

Then the rains stopped. The Llano turned brown and dusty, the flowers shriveled, the playas emptied, the bison scattered, and the grasses browned and curled down, giving a distant viewer the impression of a boundless, treeless desert. At the end of July, not only were the bison hunters tired, but also with the flowers gone and the buffalo grass dormant they could no longer smell the sweet fragrances that had resulted from the wet Llano spring.

The bison hunters did not find the Comanches. The Indian warriors and their families, convinced by Quanah, one of their leaders who had come from the reservation to get them, returned on their own to Fort Sill in Indian Territory. But the hunters got their horses back, for the Comanches had left their extra mounts behind.

With Indian people no longer on the Llano Estacado or in the Rolling Plains, bison hunters went to work in earnest. They killed hundreds of thousands of animals. During the winter of 1877–78, more than a million

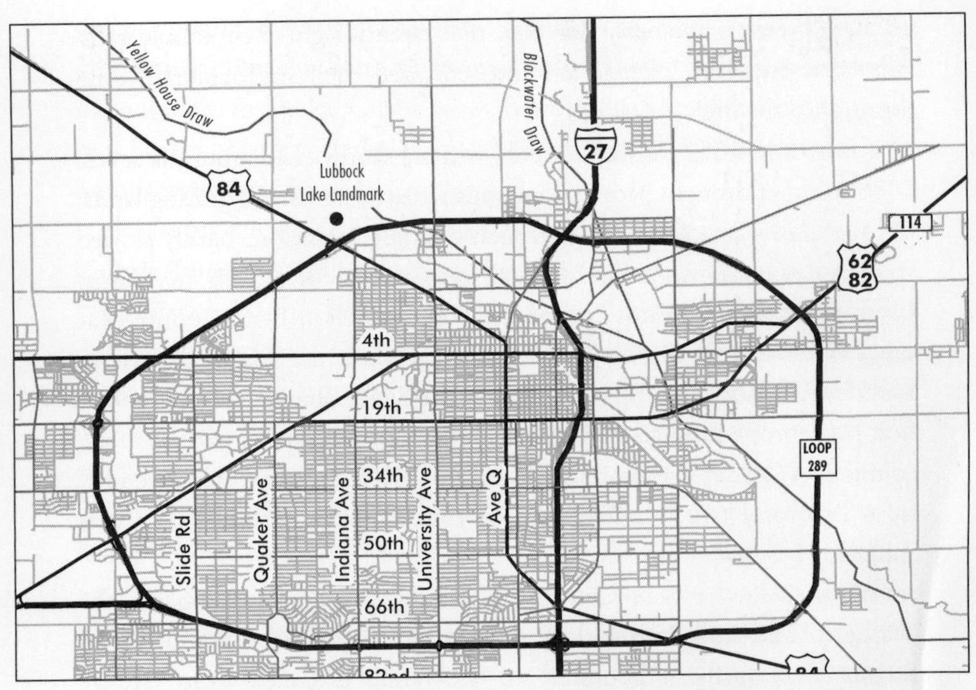

City of Lubbock Showing 19th Street & Lubbock Lake

Stadium—a site soon to be buried beneath the Marsha Sharp Freeway.

The town's roots can be traced to the Lubbock Lake site. There, before the IOA leased area pasture land, George W. Singer, a tall, long-bearded farmer-businessman who had come to the region from Ohio with a small group of Quakers, established a mercantile firm around 1883. He placed it on the southwest side of the draw, and he built a residence some distance to the west.

Locally, Singer's Store became well known, attracting travelers, cowboys, and a small but growing number of settlers. Rollie Burns remembers meeting six Apaches there. The combination trading post, saloon, and post office burned in 1886, causing its owner to relocate about a half mile downstream. In 1891, Singer, who had earlier joined promoters associated with the Monterey town site, moved his business to the new town of Lubbock.

In the summer of 1891 about one hundred people lived in Lubbock, but the town grew slowly. The IOA controlled half of Lubbock County,

Singer's Store

Around 1883, George W. Singer established a small mercantile business at the Lubbock Lake site in Yellow House Canyon. Apparently, he chose the spot because of water at the site, because farmer-stockmen were beginning to claim land across the northern half of Lubbock County, and because two small military trails crossed in the vicinity. Singer's Store catered to area cowboys, visitors along the trails, and a few of the first settlers in the region.

Singer, born in Ohio in 1844, had come to the Texas High Plains in 1879 with a group of Quaker settlers. Under the leadership of Paris Cox, the Quakers established a farming community northwest of present-day Lubbock on the Crosby-Lubbock county line. Within a year some of the Quakers had left, and Singer—who was not of the Quaker faith—sought another place to build his small business.

In 1881 Singer and his wife, Ruth Underhill, filed a claim for land in Yellow House Canyon and began plans for building a home and a business. They lived in a wagon and a tent for a time, and also may have utilized a dugout for living space. Singer hauled lumber and supplies for his house and store from Colorado City, approximately 110 miles away.

George Singer, although known as "old man Singer," was not yet forty years old when he opened his business. The store, a nondescript little building that "never did have a look to it" (according to those who saw it), did little more than provide a simple living for Singer's family. Area cowboys gathered there, as it soon became a place for poker games and good conversations. Singer, in addition to such usual supplies as dry goods, canned goods, flour, lard, sugar, hardware, gunpowder, and ammunition, provided alcohol, tobacco, and candy. He never locked the building, and often he retired to his home two hundred or so feet away while the cowboys continued their card games. It was not unusual for the cowboys to grab a piece of stick candy or a plug of tobacco and leave money on the counter for what they took.

After a fire in 1886, Singer moved his store a short distance downstream, and in 1890 or 1891 he relocated to the new community of Lubbock. George and Ruth had five children, four together and a daughter from George's first marriage. The Singers left Lubbock in 1897 and settled on a small farm in Neosho County in southeastern Kansas. Singer died in 1910.

---

making conditions difficult for farmer-settlers to secure land. Moreover, in the 1890s, drought discouraged settlement. In 1900, fewer than three hundred people lived in Lubbock County.

Growth came after the turn of the century. New state laws allowed farmer-stockmen easier access to land once dominated by large ranchers. The IOA had gone broke and began selling its property. In 1909, the

Downtown Lubbock in 1909. About a mile away the northern boundary of the former IOA Ranch (along modern 19th Street) stretched across Lubbock County. Courtesy Southwest Collection.

Santa Fe Railroad entered Lubbock, creating a boom of sorts. In addition, increased irrigation, using water from the Ogallala Aquifer, encouraged the expansion of farming, especially cotton growing. Such developments brought more people to the region.

For Lubbock, a second boom occurred in the 1920s. Texas Technological College was in part responsible. In 1925 the school opened with nine hundred students. Increasing enrollment in subsequent years sparked population growth in the area, and in 1930 Lubbock counted twenty thousand inhabitants, making it the largest city on the southern Llano Estacado.

But at the Lubbock Lake site, water disappeared. Rainfall amounts decreased in the late 1920s and early 1930s. Lubbock, for example, received only 9.59 inches of rain in 1927 and 9.72 inches in 1934. City water wells multiplied and farmers pushed down more wells to irrigate expanding cotton crops. Such a combination of factors led to a declining water table. And when the table dropped below the level of springs that for ages had provided water, the springs did not flow. In the early 1930s Lubbock Lake, which had served animals and humans for nearly twelve thousand years, shrank. It became little more than a boggy, muddy bottom land, and eventually it went dry. Grass appeared on the rich bottom, hiding evidence that a lake once existed in the canyon's sharp curve.

Meanwhile, Lubbock city officials developed plans for the lake. They had purchased ninety-three acres at the site, hoping to use the area as a future water reserve. They believed the lake's drying up had been caused by plugged springs. Not completely understanding that the water table in the area had dropped below a level necessary to supply water, they planned to unseal the springs and thereby refill the lake.

The federal government agreed to help. During the Great Depression of the 1930s unemployment rates reached over 20 percent of the population. Obviously, far too many people were without work, and the economy needed government assistance. In response the federal government under the leadership of President Franklin D. Roosevelt launched a number of emergency programs to provide work and to relieve some of the worst economic and financial distress. Collectively, the programs became known as Roosevelt's New Deal.[2]

Lubbock leaders signed on to one of the New Deal's projects: the Works Progress Administration (WPA), a program designed in part to create jobs. The city sponsored and in 1935 received a WPA project to dig in the draw, clean out the springs, and get water flowing at the lake. And, of course, the activity would put people to work.

Accordingly, in May 1936 excavation began at the Lubbock Lake site. Using picks, hoes, shovels, and wheelbarrows, a large number of men, earning forty cents per hour, cleared out the grass, dug into the earth, and hauled the dirt and mud away. As they dug the workers found that the Yellow House water table existed just below the canyon's surface. The more they dug, the more they battled with water oozing into the work area. Digging became increasingly difficult, but the men had removed about forty thousand cubic yards of material—a lot, quite frankly—when officials halted work temporarily.

Managers at the site brought in big machinery. Their men now operated draglines, at least three of them, and backhoes at the site. Using the equipment, the men removed an additional sixty thousand cubic yards of earth.

Ogallala Aquifer

The **Ogallala Aquifer** is a large underground reservoir that extends eight hundred miles through the western High Plains from Texas to South Dakota. Its maximum width, which is the region in Nebraska and Wyoming, is four hundred miles. Its thickness ranges from a thin sliver to nearly one thousand feet. In the 1980s it contained approximately three billion acre-feet of groundwater, but because humans withdraw water faster than natural forces replenish it, the aquifer's water is disappearing, particularly in Texas.

The aquifer exists in the Ogallala Formation, an unconsolidated geologic remnant consisting of huge deposits of sand, silt, and gravel washed eastward from the Rocky Mountains millions of years ago. Erosion, likewise over a million or more years, reduced the depositional zone, which once covered all of the Great Plains region, on the east. In New Mexico, the Pecos River, inching its way north through deep time from the Rio Grande, separated the Ogallala Formation from the Rocky Mountains. In Texas, the aquifer underlies the Llano Estacado and, through irrigation, supplies most of the water needs of the Llano's large agricultural economy.

Groundwater in the aquifer exists within voids (pore spaces) between sedimentary particles that compose the Ogallala Formation. In general, the water table of the aquifer parallels the topographic surface of the High Plains. That is, water in the aquifer mainly flows west to east. Because major through-going streams, such as the Canadian, Arkansas, and Republican rivers, cut the Ogallala Aquifer into distinct entities, there is little north-south flow.

Today, as indicated above, natural recharge is insufficient to make up for withdrawals from the Ogallala Aquifer. When the Ogallala extended west to the Rockies during the Pleistocene epoch (1.6 million years ago to 10,000 years ago), natural recharge was large and the water table remained high. But erosion on the eastern edge and the Pecos River on the western side, plus the onset of drier, warmer weather in the later Pleistocene (approximately 10,000 years ago), reduced the amount of recharge. Irrigation in recent years has also impacted recharge. While a few places, such as the western sandhills region of Nebraska, get a few inches of recharge per year, the average recharge throughout the aquifer zone is less than one inch per year.

When they were finished, the men had cut a roughly U-shaped trench—about 1,300 feet long and about 110 feet wide with an average depth of 18 feet—extending around the draw's natural curve. Water seeped into the trench, the old Lubbock Lake bottom, filling the cut to its rim.

Lubbock Lake, if William Curry Holden is correct, was an attractive site. Once the mud and dirt and sand had settled, the lake was clear and

Ogallala Aquifer

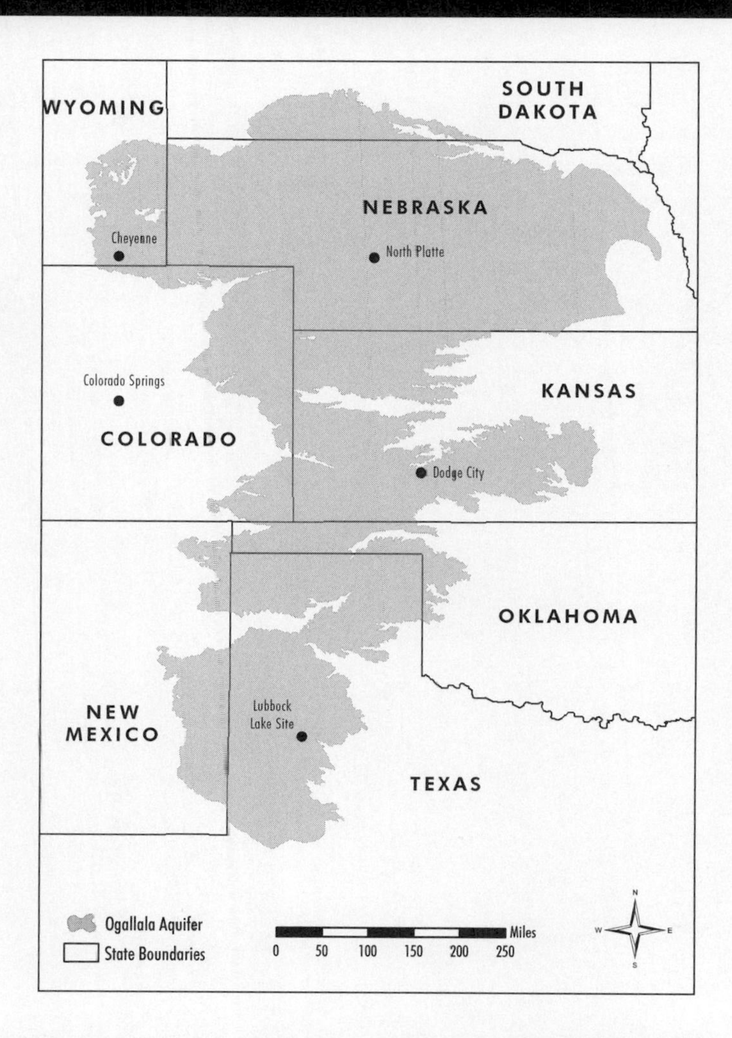

blue, and one could see the bottom. Many people visited the lake, and for some the place became a swimming hole, an alternative to the concrete pool that existed in Mackenzie Park. No lifeguards worked at the Lubbock Lake site, however, and unfortunately during the 1930s at least two boys drowned in the channel.

In the meantime, about a week after officials had completed the

Heavy irrigation, such as that represented in these photos, contributed to the declining water table that drained Lubbock Lake, or *Punta de Agua*, as the Spanish called the site. Lower photograph courtesy Southwest Collection: SWCPC 57 ZE 27.

William Curry Holden

William Curry Holden, 1896–1993, is one of the critical figures in the long struggle to preserve and investigate the Lubbock Lake site. He identified the first Folsom projectile points discovered at the site, arranged for the first systematic excavations there, and pushed hard for both protection of the site and continued archaeological excavations in the area.

Born in North Central Texas, Holden grew up on a hard scrabble farm. He received B.A., M.A., and Ph.D. degrees, the last in 1928, from the University of Texas. After teaching at McMurry College, he moved in 1929 to the very young Texas Technological College (now Texas Tech University). He played an important role in the development of the institution and in the formation of the West Texas Museum (now the Museum of Texas Tech University). He guided the growth of the Department of History and Anthropology, making Texas Tech in the late 1930s one of the few universities in North America to offer graduate work in anthropology.

Holden was instrumental in establishing Texas Tech University's Southwest Collection and the National Ranching Heritage Center. He directed archaeological fieldwork in the Texas Panhandle that helped identify the characteristic features of the pueblolike architectural style found at Antelope Creek Ruin, Saddleback Ruin, and other sites that are now assigned to the Antelope Creek phase. He also conducted anthropological fieldwork among the Yaquis in northwestern Mexico.

Although mainly a historian, Holden taught anthropology, wrote novels, and published in the field of ethnohistory and archaeology. His many books dealing with the history of West Texas and the social and economic life of the region continue to receive much-deserved attention. But his work in establishing the Museum of Texas Tech University and in preserving and developing the Lubbock Lake Landmark will be the most enduring.

William Curry Holden, more than anyone else, was responsible for the establishment of the Museum of Texas Tech University and the early excavations at the Lubbock Lake site. Courtesy Southwest Collection: SWCPC 445 E5.

Lubbock Lake site, 1967, looking north and showing its sharp bend with the Siberian elms that had been planted several years earlier as a beautification project. Courtesy Southwest Collection: SWCPC 57 (W) E3.

trench, several boys found projectile points in the excavated earth. Two of them, fourteen-year-old cousins Clark and Turner Kimmel, took their points to William Curry Holden, who chaired the Department of History and Anthropology at Texas Technological College. Holden recognized the flint projectiles as Folsom points, and hurried out to the discovery site. He and others found bison, mammoth, and horse bones.

Holden understood the archaeological importance of Lubbock Lake. But, lacking financial resources, he could not immediately undertake a systematic excavation of the site.

Lubbock Lake became a recreational area. Swimming, as noted, was the big attraction. In October of 1938, as something of a beautification effort, the Lubbock Fire Department planted trees—Siberian elms—around the place. The saplings grew rapidly and their spreading seeds soon produced an abundance of small trees, particularly down the draw.

Finally, in 1939, with a federal WPA grant, Holden got money for a

Close-up view of Lubbock Lake site, 1967, showing in detail the Siberian elms. Courtesy Southwest Collection: SWCPC 57 (W) E3.

controlled archaeological dig. He brought in Joe Ben Wheat, curator of anthropology at the University of Colorado Museum in Boulder, to supervise the excavations. With members of the West Texas Museum Association, Wheat and Holden spent several weeks late in the summer digging a couple of pits in the cut area. They abandoned the pits shortly afterward because they found no cultural material and because they encountered water. They then concentrated on digging a long trench in the southwestern edge of the draw's sharp curve. The trench, about five feet deep and more than two hundred feet long with a southwest-northeast orientation, extended from the WPA cut toward the draw's southwest wall.

Two years later—in 1941—Wheat supervised some additional work at the site. This time workers dug a pit at a point along the trench perhaps two hundred or more feet from the cut, a place well beyond where WPA workers in 1936 had deposited fill taken from the deep cut they had dug.

Wheat, Holden, and their aides produced significant information.

Lubbock Lake Stratigraphy

Scientists have identified five principal geologic levels at the Lubbock Lake site. They have numbered the basic strata oldest (1) at the bottom to youngest (5) at the top, but they cite subdivisions in each of the major strata.

At Lubbock Lake the Blanco Formation represents bedrock. Dating to the Pliocene epoch (5.1 million years ago to 1.6 million years ago), it is an extensive lacustrine (lake) deposit of dolomite and sand deposited in east-west-oriented basins cut into the Ogallala Formation, formed millions of years ago from outwash sands and gravels as the Rocky Mountains pushed up. (Of course, at other places around the Llano Estacado, such as Palo Duro Canyon, one can find older rock formations, including Permian and Triassic redbeds, which deep below the Ogallala form the rock foundation of the larger region.)

The strata at Lubbock Lake vary in depth with Strata 1 and 4 having the greatest thickness. Strata 2 and 3 are thinner. Soils identified at the site include Firstview (associated with Stratum 2), Yellow House (associated with Stratum 3), Lubbock Lake (associated with Stratum 4), Apache (associated with Stratum 5), and Singer (also associated with Stratum 5).

Stratum 1 is divided into substrata A, B, and C. Just when deposition of Stratum 1 began is unknown, but the end of the deposition was about 11,000 years ago. The bottom two substrata, 1A and 1B, represent alluvial sediments of gravel and sand deposited by running water, suggesting some years of higher precipitation. Substratum 1C is composed of sandy clay deposited during a time of waning alluviation, reflecting a smaller stream and diminished rainfall.

Stratum 2 is divided into substrata A and B. It consists primarily of lacustrine and marsh deposits. Substratum 2A contains beds of pure diatomite and peaty muds, suggesting periods of standing water. Substratum 2B consists

They recovered cultural material from the Archaic, Ceramic, Protohistoric, and Historic periods. They found three knives, scrapers, one drill, several projectile points, bones, and other materials. The trench's depth reached to the top of what local archaeologists call Stratum 3, a silty, highly calcareous, lake-deposited layer of earth known as Yellow House Soil. The stratigraphic record it provided added important data for subsequent investigations.

The water level at the Lubbock Lake reservoir also dropped. With irrigation increasing in the early 1940s and pumping from underground sources accelerating, the water table in the WPA channel fell. In fact, it

of organic silt and clay with phytoliths and diatoms, suggesting that at the time the site was an aggrading, boglike area with little or no standing water. Deposition of substratum 2A began about eleven thousand years ago and ended one thousand years later. Deposition of substratum 2B, which began about 10,000 years ago, ended about 8,500 years ago.

Stratum 3 contains lacustrine sediments of alkaline silts, clay, and loams. Deposition at Stratum 3 began about 6,300 years ago. The Yellowhouse Soil—which developed in Stratum 3—in the upper portions of the stratum represents the presence of a stable landscape and perhaps the end of middle Archaic drought conditions.

Stratum 4, divided into subtrata A and B, stands as the thickest of the five strata at Lubbock Lake. Substratum 4A contains sandy alluvial sediments deposited by an intermittent stream. Substratum 4B contains organic-rich marsh sediments, suggesting drought conditions. Deposition at substratum 4A began before 5,500 years ago and ended about 500 years later, and deposition of substratum 4B occurred during the next 500 years. The Lubbock Lake Soil developed in this stratum as deposition ended.

Stratum 5, the youngest deposit at the site, is divided into substrata A and B. Substratum 5A is clayey marsh material. Substratum 5B is marshy clays and eolian (wind-deposited) sands, but both substrata are also composed of layers of slopewash sand and gravel and windblown sand. Deposition at Stratum 5 was recent, within the last 750 years. Between periods of deposition the Apache Soil and, later, the Singer Soil formed.

Source: Vance T. Holliday and B. L. Allen, "Geology and Soils," in *Lubbock Lake: Late Quaternary Studies on the Southern High Plains*, ed. Eileen Johnson, 16-17, 19-20 (College Station: Texas A & M University Press, 1987).

dropped about two feet a year after 1939, and in 1945 the channel was dry, for the level of the water table stood below the bottom of the cut. By 1959 the water table had dropped to about thirty feet below the channel. Once again grass, weeds, and woody shrubs grew at the Lubbock Lake site.

As the reservoir dried, the stratigraphic record along the cut's almost perpendicular banks became clear. Its sharp, beautiful banding revealed five principal geologic units, or strata, and five soils. Excavations in recent years have led to significant refinement in strata and soil identification, but in a general sense scholars have numbered the strata 1 through 5 with

5 being the youngest. They have named the soils Firstview, Yellowhouse, Lubbock Lake, Apache, and Singer; Singer is the youngest.

The sharp banding along the reservoir cut encouraged new archaeological efforts. Accordingly, officials from the Texas Memorial Museum (TMM), located in Austin, led major excavations at the Lubbock Lake site in 1948, 1950, and 1951. They followed up with some additional work—sampling and testing mainly—in 1952, 1954, and 1955.

Although planned in the 1920s, the TMM building was not completed until 1938. Lack of funding during the Great Depression delayed both its establishment and its construction, but it finally opened in January 1939. The small group of historians and anthropologists who originally planned the museum hoped that it would collect, preserve, and analyze artifacts from archaeological excavations. Others wanted it to be a museum of popular exhibits, but its first director, Elias H. Sellards, undertook a vigorous research and acquisition program in geology, paleontology, and anthropology.

Accordingly, Sellards sent a team of researchers to the Lubbock Lake site. Glen L. Evans of the TMM and Grayson E. Meade from Texas Technological College supervised the investigations, which occurred at a series of places along the reservoir's walls. Unfortunately, the investigators did not write for publication much about what they found. Evans produced a description of the cut's stratigraphy based on his 1948 season at the site, and in 1952 Sellards offered a brief review of the Paleoindian material the Texas Memorial Museum efforts had uncovered, but that is the extent of it.

Nonetheless, we know that the TMM excavations were significant. The archaeologists concentrated their efforts on the Folsom-age stratum, Stratum 2, but they worked through other strata as well. They found Folsom points in situ with ancient bison bones, and in Stratum 3 they uncovered modern bison bone beds. They also discovered a stone tool from the Clovis-age stratum, Stratum 1. Perhaps just as importantly, Sellards's book on early man in North America with its review of the TMM investigations in Yellow House Canyon gave Lubbock Lake a national reputation as an important Paleoindian site.

Scientists were not the only people interested in the Lubbock Lake site.

Lubbock police personnel used an area near the site of Singer's Store for a firing range. A local Marine Corps unit practiced maneuvers along the canyon walls. The lake site served briefly as a Boy Scout camp. Other groups camped in the draw, and motorcyclists raced their machines up and down the canyon trails. And, most fearfully from an archaeologist's point of view, at a spot near the old reservoir, contractors began quarrying caliche, an activity that might have led to wholesale destruction of the site's intellectual and educational value.

In response to such activity, a small group of historians and scientists associated with Texas Technological College approached officials from the City of Lubbock. They wanted the city, which owned the area, to help protect the Lubbock Lake site from further destruction and preserve it for future excavations. Officials agreed, and in 1958 they negotiated a lease for the canyon area with Texas Tech.

In 1959 and 1960, F. Earl Green and Jane Holden Kelley directed excavations at Lubbock Lake. Green, a Slaton, Texas, native who held a Ph.D. in geology from Texas Technological College, was research assistant in vertebrate paleontology at the American Museum of Natural History in New York. In 1965 he replaced William Curry Holden as director of Tech's West Texas Museum. Kelley, who was Holden's daughter, became assistant professor of anthropology at the University of Calgary in Canada.

Founded in 1929, the West Texas Museum was a regional institution of history, anthropology, and archaeology. Short of funds in the Depression years of the 1930s, museum association members could only build the basement portion, dedicated in 1937, of their planned three-story structure. In 1950, with additional funding from Texas Tech, they completed the building. Twenty years later the renamed Museum of Texas Tech University moved into a new building. Now basically an art museum, it manages the Lubbock Lake site and operates a respected graduate program in museum science.

In the meanwhile, with a grant from the National Science Foundation and sponsored by the West Texas Museum, Green and Kelley went to work. They hoped to produce a synthesis of all previous work at the site, provide a listing of all reported artifacts from the place, and establish a clear stratigraphic record of the 1936 channel cut's wall. They conducted

West Texas Museum in the early 1950s. Now it is part of Holden Hall, which in 2005 housed the Social Science and College of Arts and Sciences offices and classrooms. Courtesy Southwest Collection: Heritage Club C 626.

most of their work on the west side of the reservoir. The results were mixed. Green prepared a long report that was never published, and Kelley produced an extensive record of known artifacts. But several archaeologists, including Kelley, questioned the correctness of Green's stratigraphy. He was wrong, they said.

Shortly after Kelly and Green finished their research, James J. Hester, professor of anthropology at the University of Colorado, sampled the Lubbock Lake site with Fred Wendorf. Wendorf, who was then with the Laboratory of Anthropology in Santa Fe, had become convinced that only through interdisciplinary efforts could a satisfactory record of human activity in the region be extracted from archaeological sites on the Llano Estacado. Therefore, he launched the High Plains Paleoecology Project (HPPP) to involve archaeology, geology, paleobotany, paleoecology, paleontology, stratigraphy, and other disciplines in a cooperative effort of investigation. The project ended in 1962 when Wendorf headed for Egypt for other research.

Before it ended, however, Wendorf, Hester, and others associated with HPPP investigated several sites on the Llano Estacado. They visited Lubbock Lake in 1961 to sample paleobotanical and paleontological materials. One of their conclusions, one that has since been shown to be incorrect, was that a coniferous forest once covered the Llano Estacado. They reached such a conclusion on the basis of pollen counts found in Paleoindian-age soils. Since then additional studies have shown that Wendorf and Hester based their findings on insufficient pollen records, counts that were too minor to prove the existence of a wet, cold boreal forest.

Data now clearly support a different view. During the Paleoindian period, the Llano Estacado was a grassland with abundant ponds and streams and with trees along the northeastern escarpment. There was no forest. The region's environment in the Clovis-Folsom years, moreover, was moving to a relatively warmer and drier climate, one that would discourage the spread of a forest ecology.

After the High Plains Paleoecology Project samplings in 1961, a decade passed without excavations at the Lubbock Lake site. Grass, weeds, shrubs, and trees grew in the draw. Rain washed gravel and mud into some of the old dig sites. Trash blew into the canyon, and some of it lodged in the shrubs and lower branches of trees, especially the Siberian elms, which, when not trimmed, grow in a profusion of spindly branches. The place became unsightly and unattractive; but in 1971 officials in Washington listed it on the National Register of Historic Places.

Then, in 1972 the Lubbock Lake Project (LLP) began. Craig C. Black, a vertebrate paleontologist at the University of Kansas, became director of the Museum of Texas Tech University. He had replaced J. Knox Jones, the interim director at the time. Black, excited by the possibilities of renewing archaeological investigations in Yellow House Draw, launched the LLP. Influenced, too, by modern analytical techniques and new methods in archaeology, he wanted, like Fred Wendorf before him, an interdisciplinary approach to excavations at the site. Accordingly, he hired Eileen Johnson, a zooarchaeologist, and Charles A. Johnson, a geoarchaeologist, to work at the museum.[3]

In 1972, LLP personnel focused their work on the Paleoindian record, particularly the Clovis and Folsom periods. Using the now standard interdisciplinary approach, they planned to study both the early Native Americans who used the Lubbock Lake site and the environment of the place at the time humans first occupied it. Their approach was detailed and systematic. They conducted fieldwork during the summer months, carefully collecting and recording materials that they found, and at the end of each field season, they sorted, analyzed, and interpreted their findings.

Personnel changed, but the work continued. Thomas W. Stafford, a geologist, for example, became the site's geoarchaeologist in 1976, and Vance T. Holliday, who had begun work at the Lubbock Lake site in 1976, replaced Stafford in 1979. Holliday, although he accepted a position at the University of Wisconsin, continued his connection with the LLP. Black left for the Los Angeles County Museum of Natural History, but Eileen Johnson stayed at the task. Indeed, she took over leadership of the Lubbock Lake Project and for more than thirty years has directed LLP work.

The Lubbock Lake Project has done more than dig through dirt in Yellow House Canyon. It has sponsored symposiums, organized international conferences, and published the results not only of its excavations but also of its conference presentations and symposium deliberations. Black, for example, put together perhaps the LLP's first major publication, *History and Prehistory of the Lubbock Lake Site* (1974), an edited work that examined several aspects of human occupation and animal and plant life in the draw. The book grew out of a symposium held at the museum.

A second symposium held in October 1975, rather than concentrating on the Lubbock Lake site, brought together Paleoindian specialists who worked at various sites in the United States and Canada. Interdisciplinary in approach, the presentations led to the book *Paleoindian Lifeways,* edited by Eileen Johnson and published in 1977 by the Museum of Texas Tech University.

Ten years later, in 1987, Johnson edited another major publication. Entitled *Lubbock Lake: Late Quaternary Studies on the Southern High Plains,* it represented the results of fifteen years of research at the Lubbock Lake site. Comprehensive, technical, and detailed, it reviews earlier investigations, corrects, or at least revises, some earlier reports, and presents the lake site in a broad, interdisciplinary fashion. Its emphasis is on the Paleoindian period.

Excavation at the Lubbock Lake Landmark. Courtesy Southwest Collection: SWCPC 57 (Z) E26.1.

Since 1990, the Lubbock Lake Project has changed direction a bit. Lately, a good portion of its work has been aimed at more recent occupations, such as the Ceramic, Protohistoric, and Historic periods, of the site. Moreover, it has spread its investigations to beyond Lubbock Lake. In recent years Eileen Johnson and her research partners, particularly Vance Holliday, have led excavations up and down Yellow House Canyon and have been investigating selected playa lakes, sand dunes, and other sites across the Llano Estacado, including Blackwater Draw.

Other changes have occurred. For example, the site's name has changed several times. For a brief period it was named the Lubbock Lake National Historic and State Archaeological Landmark. In archaeological literature it has been called the Lubbock Reservoir Site, Lubbock Locality, and Lubbock Lake site. Today it is called the Lubbock Lake Landmark preserve.

Ownership of the site has also changed. For a time it was City of Lub-

Yellow House Draw looking west toward Lubbock Lake Landmark from University Avenue before the modern Canyon Lakes Project. Courtesy Southwest Collection: 558.jpg.

bock property. The city leased the area to Texas Tech, and the university's museum managed the place. Between 1987 and 2000 the Texas Parks and Wildlife Department and the Museum of Texas Tech University jointly operated the site. Then in September of 2000, when the state's parks and wildlife department stepped aside, the site became the sole property of Texas Tech University, operated by the museum, and received its current name: Lubbock Lake Landmark preserve.

In the end, archaeological excavations at the Lubbock Lake site reveal clearly that humans have occupied the place periodically over the last twelve thousand years. They reveal, too, the comings and goings of some of the major animals, such as bison, animals forced to leave during extended droughts but which returned when rains improved their grassland forage. They reveal cultural lifeways, human camping patterns, and resource utilization.

Digging through dirt and silt and sand at the Lubbock Lake site has provided material for the study of ancient animals and early cultures. The pollen counts, botanical studies, vertebrate and invertebrate records, stratigraphy bands, artifact assemblages, soil analyses, and other scientific work has offered us a wide variety of information about the area's past. Together with other sources they have given us a nearly complete cultural and natural history record of the Lubbock Lake Landmark and, by extension, have provided an overview of the Texas High Plains through the region's long, deep past.

★

Notes

1. "Deep time" is John McPhee's term. McPhee used it to refer to a geologic time scale perhaps incomprehensible to humankind, one dating to half a billion years ago. See John McPhee, *Basin and Range* (New York: Farrar, Straus and Giroux, 1981), 127, 128.

2. The New Deal included programs for immediate financial relief; economic recovery and financial security; business, banking, industrial, and labor reform; and farm support, among many other causes. For how the New Deal affected Lubbock, see Harry S. Walker, "The Economic Development of Lubbock," in *A History of Lubbock,* ed. Lawrence L. Graves, 310–18 (Lubbock: The West Texas Museum Association, 1959–61).

3. Directors of the Texas Tech museum have included, in order, William Curry Holden, F. Earl Green, Eugene Kingman, Orlo Child (interim), J. Knox Jones (interim), Craig C. Black, James V. Reese (interim), Charles McLaughlin (interim), Leslie Drew, and others. In 2005, Gary Edson was the museum director, and Eileen Johnson led the Lubbock Lake Landmark investigations.

★

Selected Bibliography

Chapter 1—Constructing Deep Time: Putting Down Dirt

Alvarez, Walter. *T. Rex and the Crater of Doom.* Princeton: Princeton University Press, 1997.

Benton, Michael J. *When Life Nearly Died: The Greatest Mass Extinction of All Time.* London: Thames & Hudson, 2003.

Bonnichsen, Robson, and Karen L. Turnmire, eds. *Ice Age People of North America: Environments, Origins, and Adaptations.* Corvallis, OR: Center for the Study of the First Americans, 1999.

Boulter, Michael. *Extinction: Evolution and the End of Man.* New York: Columbia University Press, 2002.

Brand, John P. *Cretaceous of the Llano Estacado of Texas.* Austin: University of Texas, Bureau of Economic Geology, 1953.

Brown, Geoff, Chris Hawkesworth, and Chris Wilson, eds. *Understanding the Earth.* New York: Cambridge University Press, 1992.

Clark, Stuart. *Stars and Atoms: From the Big Bang to the Solar System.* New York: Oxford University Press, 1995.

Czerkas, Sylvia J., and Stephen A. Czerkas. *Dinosaurs: A Global View.* Rev. ed. New York: Barnes & Noble Books, 1995.

Dalrymple, G. Brent. *The Age of the Earth.* Palo Alto, CA: Stanford University Press, 1991.

Evans, Glen L., ed. *Cenozoic Geology of the Llano Estacado and Rio Grande Valley.* Lubbock: West Texas Geological Society Guidebook, 1949.

Farlow, James O., and M. K. Brett-Surman, eds. *The Complete Dinosaur.* Bloomington: Indiana University Press, 1997.

Gee, Henry. *In Search of Deep Time: Beyond the Fossil Record to a New History of Life.* Ithaca, NY: Cornell University Press, 1999.

Grambling, Jeffrey, and Barbara J. Tweksbury, eds. *Proterozoic Geology of the Southern Rocky Mountains.* Boulder, CO: The Geological Society of America, Special Paper No. 235, 1989.

Gribbin, John. *Our Changing Planet.* New York: Thomas Y. Crowell Company, 1977.

Guy, Duane F., ed. *The Story of Palo Duro Canyon.* 2nd Printing. Lubbock: Texas Tech University Press, 2001.

Hawking, Stephen W. *Brief History of Time: From the Big Bang to Black Holes.* Toronto: Bantam Books, 1988.

Holliday, Vance T. *Paleoindian Geoarchaeology of the Southern High Plains.* Austin: University of Texas Press, 1997.

Jacobs, Louis. *Lone Star Dinosaurs.* College Station: Texas A&M University Press, 1995.

Johnson, Eileen, ed. *Lubbock Lake: Late Quaternary Studies on the Southern High Plains.* College Station: Texas A&M University Press, 1987.

Kraulis, J. A. *The Rocky Mountains: Crest of a Continent.* New York: Facts on File Publications, 1986.

Kurten, Bjorn. *The Age of Mammals.* New York: Columbia University Press, 1972.

Levi-Setti, Riccardo. *Trilobites.* Chicago: University of Chicago Press, 1975.

Lockley, Martin, and Adrian P. Hunt, *Dinosaur Tracks and Other Footprints of the Western United States.* New York: Columbia University Press, 1999.

McPhee, John. *Basin and Range.* New York: Farrar, Straus and Giroux, 1981.

Nesje, Atle, and Svein Olaf Dahl. *Glaciers and Environmental Change.* London: Arnold, 2002.

Oerlemans, Johannes. *Glaciers and Climate Change.* Lisse, Netherlands: A. A. Balkema Publishers, 2001.

Oreskes, Naomi, ed. *Plate Tectonics: An Insider's History of the Modern Theory of the Earth.* Boulder, CO: Westview Press, 2001.

Padian, Kevin, ed. *The Beginning of the Age of Dinosaurs: Faunal Change Across the Triassic-Jurassic Boundary.* Cambridge: Cambridge University Press, 1986.

Parsons, Keith M. *The Great Dinosaur Controversy: A Guide to the Debates.* Santa Barbara: ABC-Clio, 2004.

Powell, James Lawrence. *Mysteries of Terra Firma: The Age and Evolution of the World.* New York: The Free Press, 2001.

Rathjen, Frederick W. *The Texas Panhandle Frontier.* Rev. ed. Lubbock: Texas Tech University Press, 1998.

Ratkevich, Ronald Paul. *Dinosaurs of the Southwest.* Albuquerque: University of New Mexico Press, 1976.

Ribeiro, Antonio. *Soft Plate and Impact Tectonics.* Heidelberg, Germany: Springer-Verlag, 2002.

Sagan, Carl. *Cosmos.* New York: Random House, 1980.

Schults, Gwen. *Ice Age Lost.* Garden City, NY: Anchor Press/Doubleday, 1974.

Stewart, Kathlyn M., and Kevin L. Seymour, eds. *Palaeoecology and Palaeoenvironments of Late Cenozoic Mammals.* Toronto: University of Toronto Press, 1996.

Szebehely, Victor G. *Adventures in Celestial Mechanics: A First Course in the Theory of Orbits.* Austin: University of Texas Press, 1989.

Ward, Peter D. *Time Machines: Scientific Exploration in Deep Time.* New York: Springer-Verlag New York, 1998.

Welty, Joel Carl. *The Life of Birds.* 3rd ed. Philadelphia: Saunders College Publishing, 1982.

Wendorf, Fred, and James J. Hester, eds. *Late Pleistocene Environments of the Southern High Plains.* Dallas: Southern Methodist University Press, Fort Burgwin Research Center, 1961.

Chapter 2—The Paleoindian Period: Hunting Big Game

Agusti, Jordi, and Mauricio Anton. *Mammoths, Sabertooths, and Hominids: 65 Million Years of Mammalian Evolution in Europe.* New York: Columbia University Press, 2002.

Arsuaga, Juan Luis. *The Neanderthal's Necklace: In Search of the First Thinkers.* New York: Four Walls Eight Windows, 2001.

Black, Craig C., ed. *History and Prehistory of the Lubbock Lake Site.* Lubbock: *The Museum Journal* 15 (1974): 1–160.

Bonnichsen, Robson, et al., eds. *Paleoamerican Origins: Moving Beyond Clovis.* College Station: Texas A&M University Press, 2005.

Bonnichsen, Robson, and Karen L. Turnmire, eds. *Clovis: Origins and Adaptation.* Corvallis, OR: Center for the Study of the First Americans, 1991.

Bonnichsen, Robson, and Karen L. Turnmire, eds. *Ice Age People of North America: Environments, Origins, and Adaptations.* Corvallis, OR: Center for the Study of the First Americans, 1999.

Bonnichsen, Robson, and Marcella Sorg, eds. *Bone Modification.* Orono, ME: Center for the Study of the First Americans, 1989.

Bradley, Robert D., et al. "Checklist of the Recent Vertebrate Fauna of the Lubbock Lake Landmark State Historical Park: 1995–1997." Lubbock: Museum of Texas Tech University, Occasional Papers, No. 184, 1998.

Brand, John P., ed. *Mesozoic and Cenozoic Geology of the Southern Llano Estacado.* Lubbock: ICASALS, Texas Tech University, 1974.

Browman, David L., and Stephen Williams, eds. *New Perspectives on the Origins of Americanist Archaeology.* Tuscaloosa: University of Alabama Press, 2002.

Collins, Michael. *Clovis Blade Technology: A Comparative Study of the Keven Davis Cache, Texas.* Austin: University of Texas Press, 1999.

Crawford, Michael H. *The Origins of Native Americans: Evidence from Anthropological Genetics.* Cambridge: Cambridge University Press, 1998.

Cummins, W. F. *Report on the Geography, Topography, and Geology of the Llano Estacado or Staked Plains.* Third Annual Report of the Geological Survey of Texas, 1891. Austin: Ben Jones & Co., 1892.

Dillehay, Thomas D. *The Settlement of the Americas: A New Prehistory.* New York: Basic Books, 2001.

Dixon, E. James. *Bones, Boats, & Bison: Archeology and the First Colonization of Western North America.* Albuquerque: University of New Mexico Press, 2000.

Dewar, Elaine. *Bones: Discovering the First Americans.* New York: Carroll & Graff Publishers, 2001.

Doherty, Mary L., and Emma Adams. *The Folsom, New Mexico, Story and Pictorial Review.* Folsom, NM: n.p., 1976.

Fagan, Brian M. *Ancient North America: The Archaeology of a Continent.* 3rd ed. New York: Thames & Hudson, 2000.

Fiedel, Stuart J. "Older than We Thought: Implications of Correct Dates for Paleoindians." *American Antiquity* 64 (January 1999): 95–116.

Fox, John W., Calvin B. Smith, and Kenneth T. Wilkins. *Proboscidean and Paleoindian Interactions.* Waco, TX: Baylor University Press, 1992.

Frison, George C. *Prehistoric Hunters of the High Plains.* New York: Academic Press, 1978.

Grayson, Donald. "Late Pleistocene Mammalian Extinctions in North America: Taxonomy, Chronology and Explanations." *Journal of World Prehistory* 5 (1991): 193–231.

Green, F. E. "Additional Notes on Prehistoric Wells at the Clovis Site." *American Antiquity* 28 (October 1963): 230–34.

Green, F. E. "The Clovis Blades: An Important Addition to the Llano Complex." *American Antiquity* 29 (October 1964): 145–65.

Guffee, Eddie. *The Plainview Site: Relocation and Archeological Investigation of a Late Paleo-Indian Kill in Hale County, Texas.* Plainview, TX: Wayland Baptist College Archeological Research Laboratory, 1979.

Haynes, Gary. *The Early Settlement of North America: The Clovis Era.* New York: Cambridge University Press, 2002.

Haynes, Vance. "Elephant Hunting in North America." *Scientific American* 214 (6) (1966): 104–12.

Holliday, Vance T. *Paleoindian Geoarchaeology of the Southern High Plains.* Austin: University of Texas Press, 1997.

Jennings, Jesse D. "Perspective." In *The Native Americans: Ethnology and Backgrounds of the North American Indians,* edited by Robert F. Spencer, et al., 6–25. New York: Harper & Row, 1977.

Johanson, Donald, and James Shreeve. *Lucy's Child: The Discovery of a Human Ancestor.* New York: William Morrow and Company, 1989.

Johnson, Eileen, ed. *Ancient Peoples and Landscapes.* Lubbock: Museum of Texas Tech University, 1995.

Johnson, Eileen, ed. *Lubbock Lake: Late Quaternary Studies on the Southern High Plains.* College Station: Texas A&M University Press, 1987.

Johnson, Eileen, ed. *Paleoindian Lifeways.* Lubbock: *The Museum Journal* 17 (1977): 1–150.

Kehoe, Alice B. *North American Indians: A Comprehensive Account.* 2nd ed. Englewood Cliffs, NJ: Prentice Hall, 1992.

Kurten, Bjorn, and Elaine Anderson. *Pleistocene Mammals of North America.* New York: Columbia University Press, 1980.

La Vere, David. *The Texas Indians.* College Station: Texas A&M University Press, 2004.

Lepper, Bradley T., and Robson Bonnichsen, eds. *New Perspectives on the First Americans.* College Station: Texas A&M University Press, 2004.

MacPhee, Ross D. E. *Extinctions in Near Time: Causes, Contexts, and Consequences.* New York: Kluwer Academic/Plenum Publishers, 1999.

Martin, Paul S. "The Discovery of America." *Science* 179 (1973): 969–74.

Martin, Paul S., and H. E. Wright, eds. *Pleistocene Extinctions: The Search for a Cause.* New Haven: Yale University Press, 1967.

Perttula, Timothy K. *The Prehistory of Texas.* College Station: Texas A&M University Press, 2004.

Rightmire, G. Philip. *The Evolution of Homo Erectus: Comparative Anatomical*

Studies of an Extinct Human Species. Cambridge: Cambridge University Press, 1990.

Roosevelt, Anna Curtenius. "Who's on First?" *Natural History* 109 (July–August 2000): 76–79.

Schultz, Gwen. *Ice Age Lost.* Garden City, NY: Anchor Press/Doubleday, 1974.

Sellards, E. H., Glen L. Evans, and Grayson E. Meade. "Fossil Bison and Associated Artifacts from Plainview, Texas." *Geological Society of America Bulletin* 58 (1947): 927–54.

Simon, Margaret. "Lubbock Lake Landmark." *Vistas: Texas Tech Research* 1 (Fall 1991): 18–21.

Spencer, Robert F., et al., eds. *The Native Americans: Ethnology and Backgrounds of the North American Indians.* New York: Harper & Row, 1977.

Stanley, Steven M. *Children of the Ice Age: How a Global Catastrophe Allowed Humans to Evolve.* New York: Harmony Books, 1996.

Stark, Michael D. "Blackwater Draw: An Archaeological Cornucopia." *Southwest Heritage* 5 (Summer 1975): 38–41.

Stewart, Kathlyn M., and Kevin L. Seymour, eds. *Palaeoecology and Palaeoenvironments of Late Cenozoic Mammals.* Toronto: University of Toronto Press, 1996.

Stringer, Christopher, and Clive Gamble. *In Search of the Neanderthals: Solving the Puzzle of Human Origins.* New York: Thames and Hudson, 1993.

Tattersall, Ian. *The Last Neanderthal: The Rise, Success, and Mysterious Extinction of Our Closest Human Relatives.* Rev. ed. New York: Nevraumont Publishing Company, 1999.

Thomas, David Hurst. *Skull Wars: Kennewick Man, Archaeology, and the Battle for Native American Identity.* New York: Basic Books, 2000.

Van Der Veen, C. J. *Fundamentals of Glacier Dynamics.* Rotterdam, Netherlands: A. A. Balkema Publishers, 1999.

Wendorf, Fred, and James J. Hester. "Early Man's Utilization of the Great Plains Environment." *American Antiquity* 28 (1962): 159–71.

Wendorf, Fred. *Paleoecology of the Llano Estacado.* Santa Fe: Museum of New Mexico Press, Fort Burgwin Research Center, 1961.

Zeuner, Frederick E. *The Pleistocene Period: Its Climate, Chronology and Faunal Successions.* London: Hutchinson & Co., 1959.

Chapter 3—The Archaic Period: Living Through Drought

Black, Craig C., ed. *History and Prehistory of the Lubbock Lake Site.* Lubbock: *The Museum Journal* 15 (1974): 1–160.

Calloway, Colin G. *One Vast Winter Count: The Native American West Before Lewis and Clark.* Lincoln: University of Nebraska Press, 2003.

Dillehay, Thomas D. "Late Quaternary Bison Population Changes on the Southern Plains." *Plains Anthropologist* 19 (August 1974): 180–96.

Dort, Wakefield, and J. Knox Jones, eds. *Pleistocene and Recent Environments of the Southern High Plains.* Lawrence: University of Kansas Press, 1970.

Fagan, Brian M. *Ancient North America: The Archaeology of a Continent.* 3rd ed. New York: Thames & Hudson, 2000.

Holliday, Vance T., ed. *Guidebook to the Central Llano Estacado.* Lubbock: ICASALS, Texas Tech University, 1983.

Holliday, Vance T., and Eileen Johnson, eds. *Fifty Years of Discovery: The Lubbock Lake Landmark, Guidebook to the Quaternary History of the Llano Estacado.* Lubbock: Museum of Texas Tech University, 1990.

Holliday, Vance T., and Eileen Johnson, eds. *Guidebook to the Quaternary History of the Llano Estacado.* Lubbock: ICASALS, Texas Tech University, 1983.

Holliday, Vance T. "Middle Holocene Drought on the Southern High Plains." *Quaternary Research* 31 (1989): 74–82.

Hopper, Kippra D. "Land of Ages." *Vistas: Texas Tech Research* 11 (Summer 2003): 12–23.

Jennings, Jesse D., ed. *Ancient North Americans.* San Francisco: W. H. Freeman and Co., 1983.

Jennings, Jesse D., and Edward Norbeck, eds. *Prehistoric Man in the New World.* Chicago: University of Chicago Press, 1964.

Jennings, Jesse D. "Perspective." In *The Native Americans: Ethnology and Backgrounds of the North American Indians,* edited by Robert F. Spencer, et al., 6–25. New York: Harper & Row, 1977.

Johnson, Eileen, ed. *Lubbock Lake: Late Quaternary Studies on the Southern High Plains.* College Station: Texas A&M University Press, 1987.

Josephy, Alvin M., Jr. *The Indian Heritage in America.* New York: Alfred A. Knopf, 1970.

Kehoe, Alice B. *North American Indians: A Comprehensive Account.* 2nd ed. Englewood Cliffs, NJ: Prentice Hall, 1992.

Kurlansky, Mark. *Cod: A Biography of the Fish That Changed the World.* New York: Walker and Company, 1997.

Martin, Calvin, ed. *The American Indian and the Problems of History.* New York: Oxford University Press, 1987.

Martin, Calvin. *Keepers of the Game: Indian-Animal Relations and the Fur Trade.* Berkeley: University of California Press, 1978.

McDonald, Jerry N. *North American Bison: Their Classification and Evolution.* Berkeley: University of California Press, 1981.

Perttula, Timothy K., ed. *The Prehistory of Texas.* College Station: Texas A&M University Press, 2004.

Quigg, J. Michael. "A Late Archaic Bison Processing Event in the Texas Panhandle." *Plains Anthropologist* 43 (1998): 367–83.

Reeves, Brian. "The Concept of an Altithermal Cultural Hiatus in Northern Plains Prehistory." *American Anthropologist* 75 (October 1973): 1221–53.

Richerson, Peter J., Robert Boyd, and Robert L. Bettinger. "Was Agriculture Impossible During the Pleistocene but Mandatory During the Holocene? A Climatic Change Hypothesis." *American Antiquity* 66 (July 2001): 387–412.

Spielmann, Katherine A., ed. *Farmers, Hunters, and Colonists: Interactions Between the Southwest and the Southern Plains.* Tucson: University of Arizona Press, 1991.

Wedel, Waldo R. *Prehistoric Man on the Great Plains.* Norman: University of Oklahoma Press, 1961.

Wendorf, Fred, and James J. Hester, eds. *Late Pleistocene Environments of the Southern High Plains.* Dallas: Southern Methodist University, Fort Burgwin Research Center, 1975.

Chapter 4—The Modern Period: Surviving Great Change

Anderson, Gary Clayton. *The Indian Southwest, 1580–1830: Ethnogenesis and Reinvention.* Norman: University of Oklahoma Press, 1999.

Barsness, Larry. *Heads, Hides & Horns: The Complete Buffalo Book.* Fort Worth: Texas Christian University Press, 1985.

Baugh, Timothy G. "Culture History and Protohistoric Societies in the Southern Plains." *Plains Anthropologist* 31 (November 1986): 167–87.

Calloway, Colin G. *One Vast Winter Count: The Native American West Before Lewis and Clark.* Lincoln: University of Nebraska Press, 2003.

Collins, M. B. "A Review of the Llano Estacado Archaeology and Ethnohistory." *Plains Anthropologist* 16 (1971): 85–104.

Elam, Earl H. "The Origin and Identity of the Wichita." *Kansas Quarterly* 3 (1971): 13–20.

Flint, Richard, and Shirley Cushing Flint, eds. *The Coronado Expedition to Tierra Nueva: The 1540–1542 Route Across the Southwest.* Boulder: University Press of Colorado, 1997.

Flores, Dan. *Caprock Canyonlands: Journeys into the Heart of the Southern Plains.* Austin: University of Texas Press, 1990.

Flores, Dan L. "Bison Ecology and Bison Diplomacy: The Southern Plains from 1800 to 1850." *Journal of American History* 78 (September 1991): 465–85.

Graves, Lawrence L., ed. *A History of Lubbock.* Lubbock: The West Texas Museum Association, 1959–61.

Habicht-Mauche, Judith A. "Coronado's Querechos and Teyas in the Archaeological Record of the Texas Panhandle." *Plains Anthropologist* 37 (1992): 247–59.

Hamalainen, Pekka. "The Rise and Fall of Plains Indian Horse Cultures." *Journal of American History* 90 (December 2003): 833–62.

Henderson, Nancy Parrott. *The Jumanos: Hunters and Traders of the South Plains.* Austin: University of Texas Press, 1994.

Henderson, Nancy Parrott. "Jumano: The Missing Link in South Plains History." *Journal of the West* 29 (October 1990): 4–12.

John, Elizabeth A. H. *Storms Brewed in Other Men's Worlds: The Confrontation of Indians, Spanish, and French in the Southwest, 1540–1795.* College Station: Texas A&M University Press, 1975.

Johnson, Eileen, ed. *Lubbock Lake: Late Quaternary Studies on the Southern High Plains.* College Station: Texas A&M University Press, 1987.

Johnson, Eileen, Vance T. Holliday, Michael J. Kaczoar, and Robert Stuckenrath. "The Garza Occupation at the Lubbock Lake Site." *Bulletin of the Texas Archaeological Society* 48 (1977): 83–109.

Kavanagh, Thomas W. *Comanche Political History: An Ethnohistorical Perspective, 1706–1875.* Lincoln: University of Nebraska Press, 1996.

Kelley, J. Charles. "Juan Sabeata and Diffusion in Aboriginal Texas." *American Anthropologist* 57 (October 1955): 981–95.

La Vere, David. *The Texas Indians.* College Station: Texas A&M University Press, 2003.

Morris, John Miller. *El Llano Estacado: Exploration and Imagination on the High Plains of Texas and New Mexico, 1536–1860.* Austin: Texas State Historical Association, 1997.

Perttula, Timothy K., ed. *The Prehistory of Texas.* College Station: Texas A&M University Press, 2004.

Pike, Albert. *Prose, Sketches, and Poems Written in the Western Country.* Edited by David J. Weber. College Station: Texas A&M University Press, 1987.

Rathjen, Frederick W. *The Texas Panhandle Frontier.* Rev. ed. Lubbock: Texas Tech University Press, 1998.

Schlesier, Karl H., ed. *Plains Indians, A.D. 500–1500: The Archaeological Past of Historic Groups.* Norman: University of Oklahoma Press, 1994.

Smith, F. Todd. *The Wichita Indians: Traders of Texas and the Southern Plains, 1540–1845.* College Station: Texas A&M University Press, 2000.

Spencer, Robert T., et al., eds. *The Native Americans: Ethnology and Backgrounds of the North American Indians.* 2nd ed. New York: Harper & Row, 1977.

Spielmann, Katherine A., ed. *Farmers, Hunters, and Colonists: Interactions Between the Southwest and the Southern Plains.* Tucson: University of Arizona Press, 1991.

Vigil, Ralph H., Frances W. Kaye, and John R. Wunder, eds. *Spain and the Plains: Myths and Realities of Spanish Exploration and Settlement on the Great Plains.* Niwit, CO: University Press of Colorado, 1994.

Wedel, Waldo R. *Prehistoric Man on the Great Plains.* Norman: University of Oklahoma Press, 1961.

Chapter 5—Reconstructing Deep Time: Digging Through Dirt

Ashburn, Percy M. *A History of the Medical Department of the United States Army.* Boston: Houghton Mifflin Co., 1929.

Black, Craig C., ed. *History and Prehistory of the Lubbock Lake Site.* Lubbock: *The Museum Journal* 15 (1974): 1–160.

Cook, John R. *The Border and the Buffalo.* Topeka, KS: Crane & Co., 1907.

Crawford, Henry B. "George W. Singer and Dry Goods Retailing on the West Texas–South Plains Frontier, 1880–1890." *West Texas Historical Association Year Book* 69 (1993): 18–33.

Crimmins, Col. M. L., ed. "Shafter's Explorations in Western Texas, 1875." *West Texas Historical Association Year Book* 9 (1933): 82–96.

Dort, Wakefield, and J. Knox Jones, eds. *Pleistocene and Recent Environments of the Southern High Plains.* Lawrence: University of Kansas Press, 1970.

Evans, Glen L., ed. *Upper Cenozoic of the Llano Estacado and Rio Grande Valley.* Lubbock: West Texas Geological Society Guidebook, 1949.

Gelo, Daniel J. "'Comanche Land and Ever Has Been': A Native Geography of the Nineteenth-Century Comancheria." *Southwestern Historical Quarterly* 103 (January 2000): 273–307.

Graves, Lawrence L., ed. *A History of Lubbock.* Lubbock: The West Texas Museum Association, 1959–61.

Holden, W. Curry. *Rollie Burns or An Account of the Ranching Industry on the South Plains.* Dallas: The Southwest Press, 1932.

Holliday, Vance T., and Eileen Johnson, eds. *Fifty Years of Discovery: The Lubbock Lake Landmark, Guidebook to the Quaternary History of the Llano Estacado.* Lubbock: Museum of Texas Tech University, 1990.

Holliday, Vance T. *Paleoindian Geoarchaeology of the Southern High Plains.* Austin: University of Texas Press, 1997.

Johnson, Eileen, ed. *Holocene Investigations at the Lubbock Lake Landmark: The 1991 Through 2000 Work.* Lubbock: Lubbock Lake Landmark Quaternary Research Center Series, No. 11, Museum of Texas Tech University, 2002.

Johnson, Eileen, ed. *Lubbock Lake: Late Quaternary Studies on the Southern High Plains.* College Station: Texas A&M University Press, 1987.

Kavanagh, Thomas W. *Comanche Political History: An Ethnohistorical Perspective, 1706–1875.* Lincoln: University of Nebraska Press, 1996.

Leuchtenburg, William E. *Franklin D. Roosevelt and the New Deal, 1932–1940.* New York: Harper, Row & Company, 1963.

Muller, William G. *The Twenty-Fourth Infantry, Past and Present.* n. p., 1928.

Murrah, David J. *C. C. Slaughter: Rancher, Banker, Baptist.* Austin: University of Texas Press, 1981.

Pike, Albert. *Prose, Sketches, and Poems Written in the Western Country.* Ed. David J. Weber. College Station: Texas A&M University Press, 1987.

Rathjen, Frederick W. *The Texas Panhandle Frontier.* Rev. ed. Lubbock: Texas Tech University Press, 1998.

Sellards, E. H. *Early Man in America: A Study in Prehistory.* Austin: University of Texas Press, 1952.

Taylor, A. J. "New Mexican Pastores and Priests in the Texas Panhandle." *Panhandle-Plains Historical Review* 56 (1984): 65–79.

Wendorf, Fred, and James J. Hester. "Early Man's Utilization of the Great Plains Environment." *American Antiquity* 28 (1962): 159–71.

West Texas Museum Association. *Half-a-Century of Progress, 1929–1979.* Lubbock: West Texas Museum Association, 1979.

★

Index